George A Evans

Hand-Book of Historical and Geographical Phthisiology

With Special Reference to the Distribution of Consumption in the United States

George A Evans

Hand-Book of Historical and Geographical Phthisiology
With Special Reference to the Distribution of Consumption in the United States

ISBN/EAN: 9783337187453

Printed in Europe, USA, Canada, Australia, Japan

Cover: Foto ©berggeist007 / pixelio.de

HAND-BOOK OF HISTORICAL AND GEOGRAPHICAL PHTHISIOLOGY

WITH SPECIAL REFERENCE TO THE DISTRIBUTION OF CONSUMPTION IN THE UNITED STATES

COMPILED AND ARRANGED BY

GEORGE A. EVANS, M. D.

MEMBER OF THE MEDICAL SOCIETY OF THE COUNTY OF KINGS, NEW YORK
MEMBER OF THE AMERICAN MEDICAL ASSOCIATION
FORMERLY PHYSICIAN TO THE ATLANTIC AVENUE, AND BUSHWICK AND EAST BROOKLYN DISPENSARIES, ETC.

NEW YORK
D. APPLETON AND COMPANY
1888

PREFACE.

In the following volume I have attempted to present a sketch of the development of our knowledge of pulmonary consumption from the time of Hippocrates up to the present day, together with the ascertained facts regarding the geographical distribution of that affection. It has also been my effort so to arrange the statistics in regard to the geographical distribution of consumption in the United States as to make them available for convenient reference in selecting localities of resort or residence for invalids, and also for those who are in health.

Dr. Bell says: "It is an almost universal practice to measure the influence of climate by the relation which different regions and localities hold to pulmonary consumption; a disease which, probably more than any other, depends upon preventable conditions, intimately associated with foul soil, or density of population." *

This treatise is made up, to a great extent, of the observations of others, and for the most part in their own

* "Climatology," etc., by A. N. Bell, A. M., M. D., New York, 1885.

words; however, I have endeavored to give full credit in all cases.

The historical portion of this work, from Hippocrates up to and including Empis, was translated almost entirely from Waldenburg's "Die Tuberculose, etc.," "Nach historischen und experimentellen Studien," Berlin, 1869. Some assistance was obtained, however, from Ruehle's "Historical Sketch" of pulmonary consumption.*

Hirsch's "Handbook of Historical and Geographical Pathology" has supplied most of the data pertaining to the geographical distribution of consumption in countries other than the United States, and his observations and opinions concerning the influence of the various factors of climate, etc., have been utilized and adopted so far as they have seemed to agree with well-established facts.

The statistical data, under the head of "Summary for States, Groups, Cities, and for Counties of Ten Thousand Population, etc., showing the Number of Deaths from Consumption per One Thousand Inhabitants," as well as other data relative to the geographical distribution of consumption in the United States, were compiled, after much labor, from the "Tenth U. S. Census Reports."

In reference to these reports, Dr. J. S. Billings says: "The term 'consumption,' as used in the enumerators' (census) returns, is, no doubt, a vague one, and includes many cases which are not due to true tubercular phthisis, especially in infants; yet it is probable that a very large

* "Ziemssen's Cyclopædia," etc., American edition.

majority of the cases thus reported are rightfully named, and that some conclusions may be drawn from the figures as to the relative prevalence of tubercular lung-disease which will be reliable to a great extent."*

At all events, they represent the most reliable data attainable at the present day.

G. A. Evans.

909 Bedford Avenue, Brooklyn, N. Y.,
May, 1888.

* "Tenth U. S. Census Report," vol. xii.

CONTENTS.

I.

HISTORICAL SKETCH.*

So far as our information goes, pulmonary consumption has always existed. "It is," as Prof. Hirsch remarks, "emphatically a disease of all times, all countries, and all races. No climate, no latitude, no occupation, no combination of favoring circumstances, forms an infallible safeguard against the onset of tuberculosis, however such conditions may mitigate its ravages or retard its progress. Like typhoid fever, phthisis dogs the steps of man wherever he may be found, and claims its victims among every age, class, and race." †

Hippocrates (460–377 B. C.) seems to have been the first to describe phthisis with any degree of clearness; he considered the affection to consist of a suppuration of the lungs due to various causes, and that it may assume an acute or chronic character.

It may occur as a result of inflammation.

It may proceed from a chronic pneumonia, which is complicated by a defluxion of mucus from the brain into the lungs.

---

* With the special assistance of "Waldenburg's "Die Tuberculose," etc., Berlin, 1869.

† "British Medical Journal."

It may arise from an extravasation of blood into the lungs, which, through its failure of absorption, is converted into pus, or it may occur whenever a collection of mucus, blood, or any morbid products in the lungs or pleural cavities, fail to be expectorated or absorbed within a specified time.

Hippocrates believed phthisis to be curable when due to either of the above causes.

He considered the affection to be non-specific, and that it always occurred as a natural result when mucus, blood, or morbid products in the lungs or pleural cavities, were converted into pus through their failure of absorption.

He recognized, however, another form of phthisis, the result of "phymata." * "Phyma" being generally translated "tuberculum" in Latin, this circumstance constitutes the only evidence that Hippocrates recognized the existence of tubercle. "Phymata" referred, according to Virchow, † to points of cheesy matter or small collections of pus. Hippocrates wrote concerning "phymata" not only of the lung, but also of the pleura, tonsil, bladder, urethra, and as a cause of fistula, so there can be scarcely any doubt that he referred to simple abscesses. In short, throughout all his writings we find that "phyma" signifies a collection of pus the result of inflammatory action, which has depended for its origin upon a failure of mucus, blood, or bile to be absorbed. In none of his works, however, does there

---

* Waldenburg. † Ibid., *loc. cit.*

exist the least evidence for supposing that tubercle of these times was known to him. It is quite evident that the only distinction Hippocrates made between "phyma" and simple suppuration was, that the former represents a circumscribed condition, while the latter is more diffused in character; however, he considered phthisis to exist when either form appears in the lungs.

Hippocrates undoubtedly found tuberculous nodules, not only in the lungs of man, but also in those of the lower animals; but he seems to have attached no importance to their existence, except to consider them as centers of simple inflammation or suppuration.

His book of "Epidemics" treats of phthisis as an epidemic disease supervening upon attacks of semi-tertian.*

In Adams's translation of Hippocrates's works, the following history of phthisis is given: "Early in the beginning of spring, and through the summer, and toward winter, many of those who had been long gradually declining, took to bed with symptoms of phthisis; in many cases, formerly of a doubtful character, the disease then became confirmed, in these the constitution inclined to the phthisical. Many, and in fact most, of them died. . . . The greatest and most dangerous disease, and the one that proved fatal to the greatest

* Francis Adams's "Genuine Works of Hippocrates," London, England. William Wood & Co., New York, 1886.

number, was consumption. With many persons it commenced during the winter, and of these some were confined to bed, and others bore up on foot; the most of those died early in spring who were confined to bed; of the others, the cough left not a single person, but it became milder through the summer; during the autumn, all these were confined to bed, and many of them died; but in the greater number of cases the disease was long protracted. Most of these were suddenly attacked with these diseases, having frequent rigors, often continual and acute fevers; unseasonable, copious, and cold sweats throughout; great coldness from which they had great difficulty in being restored to heat; the bowels variously constipated, and again immediately in a loose state, but toward the termination in all cases with violent looseness of the bowels; a determination downward of all matters collected about the lungs; urine excessive, and not good; troublesome melting. The coughs throughout were frequent, and sputa copious, digested, and liquid, but not brought up with much pain; and even when they had some slight pain, in all cases the purging of the matters about the lungs went on mildly. The fauces were not very irritable, nor were they troubled with any saltish humors; but they were viscid, white, liquid, frothy, and copious defluxions from the head. But by far the greatest mischief attending these and the other complaints was the aversion to food, as has been described. . . . The form of body peculiarly subject to phthisical complaints was the smooth, the whitish, that resembling the lentil; the reddish, the blue-eyed, the leucophlegmatic, and that with the scapulæ having the appearance of wings." ("Aphorisms." *)

"Phthisis most commonly occurs between the ages of eighteen and thirty-five years.

"In pleuritic affections, when the disease is not purged off in fourteen days, it usually terminates in empyema.

"Persons who escape an attack of quinsy, and when the disease is turned upon the lungs, die in seven days; or if they pass these they become affected with empyema.

"In persons affected with phthisis, if the sputa which they cough up have a heavy smell when poured upon coals, and if the hairs of the head fall off, the case will prove fatal.

"Phthisical persons, the hairs of whose head fall off, die if diarrhœa set in.

"In persons who cough frothy blood, the discharge of it comes from the lungs.

"Diarrhœa attacking a person affected with phthisis is a mortal symptom.

"Persons who become affected with empyema after pleurisy, if they get clear of it in forty days from the breaking of it, escape the disease; but, if not, it passes into phthisis."

Adams, in a note to the above aphorisms, states that Hippocrates applied the term empyema to the purulent expectoration that follows inflammation of the lungs and

* Adams's trans., "Genuine Works of Hippocrates."

pleurisy, and also to that which proceeds from a cavity of the lungs in tubercular phthisis.

Isocrates, a contemporary of Hippocrates, considered pulmonary phthisis to be a contagious disease; this opinion, he claimed, was based on clinical observation.

Celsus (about 30 B. C. to 50 A. D.) recognized three forms of consumption: an atrophy of lung, the result of its imperfect nourishment; cachexia, i. e., constitutional weakness, the result of protracted sickness, depressing therapeutics, imperfect nutritive processes, etc.; and ulceration of lung, which is characterized by frequent cough, putrid expectoration, and high fever.

Celsus recognized neither phyma nor tubercle in the lungs, although in the fifth volume of his work he speaks of phyma of the skin; phyma he, however, translated as tubercle, and used the term tubercle to designate tumors which occur in various pathological processes ("Furunculus vero est tuberculum acutum, etc.").

Aretæus Cappadox (50 A. D.) was the first clearly to describe pulmonary phthisis as a special pathological process, although he admitted that the pus of empyema frequently infiltrates the lungs, and gives rise to a pathological condition the symptoms of which resemble those of phthisis. Nevertheless, he drew a sharp distinction between the two affections. He considered phthisis to be due to abscess of lung, chronic bronchitis, or pulmonary hæmorrhage, and that either of these might give rise to an accumulation of pus in the lungs. The term phyma was not used by Aretæus. In the second book of his work ("De causis et signis acutorum morborum"),

Chapter I, he states that peri-pneumonia may, in its course, develop abscesses and tumors in the lungs, and that when these ulcerate they give rise to phthisis. The development of phthisis from pulmonary hæmorrhage, pleurisy, and empyema, was considered to occur only when their morbid products failed to be absorbed, were converted into pus, and finally established purulent infiltration of lung-substance. Suppuration of lung and phthisis were synonymous; however, Aretæus believed that phthisis frequently occurred as a result of chronic catarrhal bronchitis.

Galen (131–201 A. D.) ("De Methodo Mendi") describes phthisis as consisting of simple ulceration or suppuration of the lungs, by which portions of the organ are sloughed off, become putrid, and are discharged in the expectoration; he considered ulceration of the lungs in connection with ulcerations of other organs, viz., stomach, bladder, uterus, mouth, etc., and we find nowhere in his writings any recognition of phthisis as a specific disease. He believed the affection to be due chiefly to mechanical irritation of lung-tissue, frequently induced by violent respiratory action, catarrh, etc., followed by pulmonary hæmorrhage and finally ulceration. Injury to lung-tissue was considered by Galen to run the same course as injury to the tissue of other organs of the body, and that ulceration would occur if the reparative process failed to be completed in a few days. He also recognized that ulceration of the lungs might occur independent of pulmonary hæmorrhage, as a result of corrupt secretions; this form of phthisis, however, he regarded as incurable.

In order to cure pulmonary consumption Galen considered it necessary to make use of the same method of treatment which had been found to be most successful in healing ulcers in other organs of the body, the skin, stomach, bladder, uterus, etc. This treatment, he explains, consists of measures for drying up the secretions or discharges, thereby inducing cicatrization. In deference to this opinion, Galen was in the habit of sending his phthisical patients to dry-air localities. He considered the phymata of Hippocrates to be latent abscesses, which in the beginning produced no symptoms other than dyspnœa; he seems to have attached no special importance to simple abscess of the lungs, and did not consider that it constituted in itself pulmonary phthisis, although he undoubtedly believed that it might lead to the development of that affection through its irritation of lung-tissue, extensive suppuration, hæmorrhage, or blood-poisoning.

It is quite evident that Galen not only had no knowledge of tubercle of later times, but we may feel certain that he failed to recognize nodes or nodules in the lungs.

Very little advance was made in the knowledge of pulmonary phthisis from the time of Galen until the beginning of the seventeenth century; the principal writers during that interval were, according to Waldenburg, Rhazes, Maimonides, and Benedictus.

Rhazes (Al Razi), of the Arabian school (923 or 932 A. D.), adopted the opinions of Hippocrates and Galen concerning phthisis, and copied extensively from their works. In his writings on the subject he states that suppuration of the lungs may result from peri-pneumonia, pleuritis, hæmoptysis, or an injury, and that patients die from it because the lungs can not be treated like external parts by the knife or cautery.

Maimonides (1135–1204) devoted his attention chiefly to the pathological anatomy of those animals which the Jews of his time slaughtered and used as food; and, although his studies in this direction were actuated by a sense of religious duty (being a Jew), nevertheless he utilized the information thus gained in his consideration of the morbid changes which occur in the tissues of man. Maimonides made, however, no addition to the knowledge of phthisis of his time. He adopted almost entirely the opinions of Hippocrates and Galen; and, although he undoubtedly found nodules in the lungs of beef-cattle, he nevertheless failed to consider them in connection with phthisis (pearl-disease), or to attach to them any special importance.

Alexander Benedictus (1525) considered Hippocrates as the highest authority on pulmonary phthisis, and adopted his opinions concerning its etiology and pathology almost without reservation. He makes no mention of "phyma" or tubercle in his writings.

Franciscus Delevoe Sylvius (1614–1672) seems to have been the first to recognize the existence of nodes in connection with ulcerations and suppurations of the lungs. He believed in inherited or acquired predisposition to the development of phthisis, and speaks of enlarged pulmonary glands in connection with nodes, from the softening of which tubercles, both large and small, as well as cavities, are formed. He recognized two forms of pulmonary phthisis: the first, due to purulent infiltration of the lungs, resulting from hæmoptysis, peri-pneumonia, or empyema, and characterized by ulceration, suppuration, and loss of lung-substance; the second variety he considered as occurring in scrofulous subjects, and due to nodes in the lungs which suppurate to form cavities, in association with enlarged pulmonary glands which soften and are converted into tubercles.

Sylvius declared ("Tractus de Phthisi") his observations and teachings concerning pulmonary phthisis to be entirely new, and we are compelled to believe him when we consider that he was the first, so far as we can learn, to describe ulceration of the lungs as due to a suppuration of tubercles. He recognized hard, tuberculous masses in the lungs which first soften in the center, become abscesses, and finally disintegrate for the production of cavities and putrid expectoration; he believed in the existence of two kinds of tubercle, large ("tubercular majora") and small ("tubercular minora"), and it is not improbable that he recognized as "tubercular minora" the miliary tubercle of to-day.

He described nodes in the lungs as "glandulosa tubercula," and believed that invisible glands exist in the lungs as well as in other organs of the body, which, in scrofulous subjects, enlarge to form small tubercles; that these in turn develop into a larger variety, which, when they have finally attained a certain size, suppurate to form cavities, etc.—scrofula of the lungs.

Willis (1622–1675) endeavored in his writings to disprove the identity of phthisis with ulceration of the lungs. He claimed that extensive post-mortem investigations showed the affection to consist of an infiltration or hardening of lung-substance, due to imperfect or vicarious nutritive processes, characterized by the presence of tubercle, and resulting in destruction of lung-tissue, and demonstrated at the same time that ulceration of the lungs did not occur in coexistence with pulmonary phthisis. Willis undoubtedly recognized miliary tubercle.

After Willis, the writings of Bonnet (1620–1689) attract attention.

The first volume of his work on pathological anatomy gives his observations concerning over one hundred and fifty cases of pulmonary phthisis. He considered the affection due to various pathological changes in the lungs, such as ulceration, abscess, suppuration, empyema, induration, scirrhus, tubercle, etc. In certain cases he identifies tubercle with the phyma of Hippocrates, and speaks of tubercle or abscess of the pleura as "tubercula glandulosa," which shows, according to Waldenburg, that tubercle was regarded by him with the same significance as it was by Sylvius and Willis.

Bonnet made no distinction between scirrhus and tubercle. Vomicæ he considered as slowly developing abscesses, generally due to the softening or breaking down of tubercles.

Manget (1700), in a revised edition of Bonnet's works, recorded his own observations in forty-nine cases of general miliary tuberculosis. In one case (a young man who died from phthisis) he made a post-mortem examination, and found miliary tubercles ("grandines") in the lungs, liver, spleen, kidneys, mesenteric glands, and intestines. He likened these bodies to millet-seed, and considered them to be scrofulous in their nature. He states that they are found not only in the tissues of man, but also in those of the lower animals, and when they occur in the lungs they soften at first in the center, for the production of "vomicæ," which finally suppurate and develop phthisis.

Morton (1689), in his work on "Phthisiology," distinguishes different forms of pulmonary phthisis from one another—those which are due to syphilis, pneumonia, hæmoptysis, and scrofula. He believed that a general febrile condition of the system, due to impaired vitality, gives rise to the formation of nodes in the lungs in every variety of phthisis, and that their suppuration leads to a general destruction of lung-tissue. He identified nodes with tubercles, and believed that no form of pulmonary phthisis could develop without them. Morton's writings do not show that he recognized miliary tubercle.

Sydenham (1624–1689), Leigh (1694), Hoffmann (1660–1742), Boerhaave (1668–1738), Mead (1673–1754), Morgagni (1682–1771), Van Swieten (1700–1772), Sauvages (1706–1767), Anenbrugger (1722–1809), and many other prominent observers wrote concerning pulmonary phthisis. However, none of them seem to have advanced the knowledge of the subject.

Stark, whose observations and writings on phthisis pulmonalis were first published in 1785 (fifteen years after his death), gave a more accurate description of tubercles than had ever been given before, and showed how cavities were formed from them.

Reid (1785), who published the researches of Stark, together with his own, considered that tubercles were derived from coagulated lymph, and strongly opposed the old doctrine of their glandular or scrofulous nature.

Cullen (1709–1790) believed phthisis may arise from an expectoration of blood; from a suppuration of lung-tissue, the result of inflammatory action; from catarrh occurring in flat-chested subjects; and, finally, when nodes are developed in the lungs. He described small bodies (tubercles) in the lungs, resembling hardened glands, which, when they become inflamed, ulcerate and develop phthisis. Kortum (1786) occupied in his opinions concerning pulmonary phthisis a position midway between the old and new doctrines. Baume (1795) failed to consider the teachings of Stark and Reid. Either they were unknown to him, or else he regarded them as of no importance. However, Baume failed to advance the knowledge of phthisis. He believed that tubercles are developed from pulmonary glands in scrofulous subjects, and that their suppuration gives rise to phthisis.

Matthew Baillie published (in 1793) a small work entitled "The Morbid Anatomy of some of the most Important Parts of the Human Body." It made an era in medical science. In this work Baillie describes, as the most frequent lesion in the diseased (phthisical) lungs, the presence of nodes, which are at first about the size of the head of a pin, but afterward, by the coalescence of several, increase to larger nodes. The breaking down of these nodes into pus he regards as the cause of consumption. He also distinguishes them from glands. At the same time, however, he speaks of the more diffuse deposits as composed of scrofulous matter, although he thinks that they consist of the same substance as the nodes. In the lymphatic glands he speaks of this matter as "caseous." In many other organs, such as the kidneys, bladder, testicles, etc., this scrofulous matter may be present as well as the tubercles, and everywhere they both possess the common property of being converted into a soft, caseous mass.

Portal (1780) followed quite closely in the footsteps of Baillie in his opinions concerning the morbid changes which occur in the lungs in consumption. In his writings, he designates caseous matter as *tuberculous*, and says that this is the term in general use.

Portal described bronchial and lymphatic glands, and believed that tubercles were frequently derived from the latter. He states, however, that tubercles may develop in the connective tissue of the lungs, independent of these glands, through extravasation of lymph from the lymphatic vessels. He recognized not only tuberculosis of the lungs, but also of the pleura, liver, spleen, mesentery, etc., and considered tuberculosis of all these organs as hereditary scrofulous affections.

Vetter, whose "Aphorisms on Pathological Anatomy" were published in Vienna in 1803, distinguished the "phthisis pulmonalis" ("ulcus pulmonum"), which is characterized by suppuration ("vomica"), the result of an inflammation of lung-tissue, from that form of the affection whose primary lesion consists in a formation of tubercles in the lungs. He considered tubercles to be non-scrofulous in their nature, although he believed that a predisposition to their development in the lungs might be acquired as a result of external debilitating influences. He denied the glandular origin of tubercles, and considered that they develop primarily either in the openings of the small bronchial tubes or in their contiguous connective tissue.

Vetter also described tuberculosis of the peritonæum, bowels, liver, spleen, and even the uterus; however, he failed to consider tuberculosis of these organs in connection with tuberculosis of the lungs.

Bayle (1774–1816), who, according to Waldenburg, is the real founder of our knowledge of the tubercle, used the term miliary tubercle, and described a granular as well as a tubercular phthisis. He considered the former variety as not at all infrequent, although other writers do not mention it. He described six varieties of phthisis: (1), "phthisie tuberculeuse"; (2), "phthisie granuleuse"; (3), "phthisie avec melanose"; (4), "phthisie ulcereuse"; (5), "phthisie calculeuse"; (6), "phthisie cancereuse." Out of 900 cases of phthisis reported by him, 624 belonged to the first, 183 to the second, 70 to the third, 14 to the fourth, 4 to the fifth, and 3 to the sixth variety. He declared tubercular phthisis to be a distinctly specific disease, which may or may not be complicated with inflammations, catarrh, hæmoptyses, etc., but does not originate in them.

Bayle found miliary tubercles in various organs of the body, and was the first to recognize them in the larynx and trachea. He considered tubercular phthisis to be a constitutional rather than a local disease, the result of cachexia.

Hufeland (1819), who made so many important contributions to the medical literature of his day, failed to contribute anything of value to the study of phthisis.

Laennec (1781–1826). Rindfleisch says: * "It is well known that there was a time when pathological anatomy pointed with pride to its knowledge of phthisis tuberculosa. Laennec's theses concerning the gray granulation and its change into yellow tubercle spread, after long controversy, a welcome light over the nature of the entire process. The manifest resemblance of the lesions in different organs, especially in the lungs, kidneys, and mucous membranes, were explained by the axiom that there was only one phthisis—a phthisis tuberculosa."

Laennec amplified and perfected the doctrine of phthisis tuberculosa which Bayle was the first to promulgate.

† "He begins his anatomical description of phthisis with the statement that the tuberculous matter is developed under two principal forms, that of isolated bodies and that of infiltration. Both of these forms present several varieties, according to the stages of development.

---

* Ziemssen's "Cyclopædia of the Practice of Medicine," vol. v, American edition, William Wood & Co., New York, 1875.

† Ruehle, in Ziemssen's "Cyclopædia," etc.

The isolated tubercles have four principal varieties, the miliary, the crude, the granular, and the encysted tubercle; while the infiltrated has three, the irregular, the gray, and the yellow. In both forms the tuberculous matter is at first gray and hyaline, gradually becoming opaque and very dense; afterward it softens, becomes more and more fluid like pus, and is finally discharged through the bronchi, thus giving rise to tuberculous cavities. Judging from the only signs of inflammation which were then attainable, Laennec denied the inflammatory nature of tuberculous matter, and particularly that pneumonia could pass into tuberculosis. He was, moreover, just as skeptical in regard to the causation of tuberculosis by bronchial catarrh, for the reason that he was unable to satisfy himself from anatomical evidence that the latter was ever directly converted into the former."

Bichat (1771–1802) and Beclard (1821) adopted Laennec's views almost completely.

Broussais (1772–1838), Gendrin (1826), Lobstein (1829), and Lombard (1834), strongly opposed the teachings of Laennec.

Broussais considered pulmonary phthisis as chronic pneumonia, the result of irritation and inflammation of the connective tissue of the lungs. Inflammation of the pleura and bronchial catarrh also frequently give rise to phthisis. Tubercle and tuberculous matter are simply products of inflammation.

Louis (1825) adopted Laennec's views completely, and his monograph, "Recherches anatomiques, pathologiques et therapeutiques sur la Phthisie," was for a long time the standard text-book on the subject.

Andral (1842) considered tubercle to consist of a yellowish-white substance, a product of secretion, which, originally liquid, becomes friable or cheesy, and maintains this consistency because liquid between its molecules prevents their complete cohesion—the scrofulous, tuberculous, steatomatous, or cheesy matter of other writers. He speaks of "tuberculization of pus," and its conversion into cheesy matter. Although Andral opposed Laennec's views on many points, he subsequently adopted the opinion that tubercles were produced independently of any irritative or inflammatory process, but their presence excited a secondary inflammation, which ultimately expelled the tubercles.

Waldenburg speaks of Andral as the predecessor of Reinhardt and Virchow.

Meckel (1818) and Neumann (1822) identified tubercle with scrofula.

Schönlein (1839) considered phthisis a secondary disease, a result either of inflammatory or non-inflammatory changes in the lungs, independent of tuberculosis and scrofula, although it frequently occurs as a sequel to them.

Engel (1845) separated the miliary tubercle from the infiltrating tubercle; he considered the latter to be an inflammatory exudation.

Vogel (1845) derived tubercle from liquid exudation, which he considered to be a result of qualitative changes in the constituents of the blood, and of hyperinosis. Vogel made no distinction between isolated and infiltrating tubercles; he, however, distinguished tuberculosis from scrofula.

Rokitansky * (1842–1861), "whose pathological anatomy made its first appearance in 1842, declared that tubercles are neoplasms, and adopted them for both of Laennec's two forms. As late as 1861 he speaks of the miliary tubercle and the tuberculous infiltration as the two forms of tubercle. The latter, he says, consists in the impaction of the texture of the lungs with a reddish, grayish-red, grayish, finely granular, stiff, tuberculous mass, sometimes involving a whole lobe, as a lobar tuberculous infiltration, but very often lobular—that is, affecting single lobules or small aggregations of the same. The tuberculous infiltration differs from the tuberculous granulation in the fact that in the former the tuberculous substance is produced uniformly, and in so solid a form that the pulmonary structure over a large extent becomes unrecognizable and impermeable."

Alison (1824). In England, the doctrine of tuberculosis slowly developed; at first Baillie's views were generally adopted, and scrofula occupied a position in the foreground. Phthisis was regarded as a local expression of scrofula and tubercle, one of its products. Alison and many other prominent observers of his time adopted these views.

Baron (1828) identified tubercle with hydatid cysts, while Addison considered tubercle as abnormal epithelial cells which he derived from white blood corpuscles.

* Ruehle, in Ziemssen's "Cyclopædia," etc.

Carswell regarded miliary tubercle as a neoplasm, readily disposed to caseation, in consequence of which change the tubercle becomes opaque, white, and finally yellow. He defined the tuberculous infiltration as an inflammation which becomes modified in various directions, and considered caseous pneumonia to be a scrofulous affection.

Clark adopted the views of Laennec almost completely, while Stokes occupied a position, in his opinions regarding pulmonary phthisis, about midway between those which were held by Laennec and Broussais.

Gluge (1841) seems to have been the first to utilize the microscope for the investigation of tubercle.

Lebert * (1844). "The views of the French writers in regard to the specific nature of phthisis were confirmed, microscopically, by Lebert, who demonstrated small, irregularly oval, granular corpuscles, to which he gave the name tubercle corpuscles, and which he regarded as characteristic of all tuberculous matter, including both the miliary tubercle and the tuberculous infiltration."

"Reinhardt † (1847) demonstrated that tubercle corpuscles may originate from pus-cells, and thereby deprive them of their importance. By 1850 he had established the fact that many substances hitherto regarded as tubercle were identical with the products of inflammation."

Virchow (1850) founded, according to "Rind-

* Ruehle, in Ziemssen's "Cyclopædia," etc. † Ibid.

fleisch,"* "the new doctrine to supplant Laennec's teachings. He taught that only the miliary tubercles were to be called tubercles, and that no process was to be called tubercular unless the gray miliary granulations were found. Cheesy conditions could be formed from thickened pus and other cellular new growths just as well as from miliary tubercles. It should be the office of pathological anatomy to separate the cheesy products of inflammation from cheesy degeneration of miliary tubercles. . . . Virchow had called attention to the fact that, in almost all cases of acute, disseminated miliary tuberculosis, cheesy forms could be found somewhere in the body, usually a cheesy lymphatic gland. Practical medicine, however, for a long time regarded Virchow's teachings with distrust. Felix Niemeyer was the only clinical teacher who boldly adopted the new doctrine."

Ruehle † says: "All the recent numerous investigations have been based upon this histological distinction between the miliary tubercle and the infiltration; but it is not too much to say that as yet they have failed to establish conclusions which have met with general acceptance."

"Bayle's miliary tubercle plays, however, only a subordinate *rôle* in pulmonary consumption; it is an accidental secondary product. When it forms the only anatomical lesion, we have to deal with an acute infectious disease, the acute miliary tuberculosis, which does not belong to phthisis. There is probably no chronic miliary

---

* Rindfleisch, ibid. † Ruehle, in Ziemssen's "Cyclopædia," etc.

tuberculosis in the old sense of the term. Phthisis is also anatomically a chronic inflammatory disease, with intercurrent simple forms of inflammation which heal by cicatrization. But the pernicious form of phthisis is a specific variety of inflammation with characteristic caseous metamorphosis; this inflammation is localized in different parts of the tissues, is characterized by the fact that it begins with and also produces the true histological miliary tubercle of the smallest kind, and in itself undergoes no other metamorphosis except necrosis."

Lorain, Robin, and Empis denied the identity of miliary granulations with tubercle. Empis, who made extensive microscopic investigations of tubercle, considered the granulations, which Laennec derived from miliary tubercle, as a result of inflammatory action. He admitted, however, that they may occur in combination. He, as well as Lorain and Robin, strenuously opposed the doctrine which Virchow had promulgated.

Klencke * (1843): "Isolated, imperfect attempts at the artificial production of tuberculosis, made at the end of the last century, gave only negative results. The credit of the first successful experiments belongs to Klencke, who in the year 1843 succeeded in inducing an extensive tuberculosis of the lungs and liver in rabbits, by inoculation with portions of miliary and infiltrating tubercles from man, and he did this by the introduction of these masses into the veins of the neck. He did not continue

* "Investigation of Pathogenic Organisms," vol. i, by Dr. Robert Koch, 1881, translated by Victor Horsley, B. S., F. R. C. S., New Sydenham Society, London, England, 1886.

his researches, and they were consequently soon forgotten." In 1857 Buhl pointed out that an outbreak of tuberculosis was almost always attributable to the previous existence of caseous matter somewhere in the body."

Villemin presented to the French Academy of Medicine, December 4, 1865, his first memoir on the origin and nature of tubercle, and its transmission to rabbits from man. It contained a detailed record of methodical and thorough experimental investigations into the nature of tuberculosis.

"Villemin * inoculated not only with tubercular material from human beings, but also from cases of bovine tuberculosis, and proved experimentally the identity of the latter disease with tuberculosis."

Villemin's researches, from the number of his experiments, the careful manner in which they were carried out, and the employment of suitable control experiments, appeared to have decided the question in favor of the infective theory. The numerous workers, however, who repeated Villemin's experiments after the same or a modified method, arrived at very contradictory results." Villemin's conclusions were: 1. Tuberculosis is a specific affection. 2. It is produced by an infective agent. 3. Rabbits may be successfully inoculated from man. 4. Tuberculosis should be classed with virulent diseases, such as variola, scarlatina, syphilis, and glanders.

---

* Koch, ibid.

"Villemin's* conclusions were soon confirmed by a series of experiments carried on by Simon, Marcet, Clark, and Lebert. In 1868 Sanderson and Fox succeeded in producing tuberculosis in Guinea-pigs, not only by the insertion of tuberculous material, but also by that of non-tuberculous. Fox used the following non-tuberculous materials: Putrid muscles, pus of various kinds, pneumonia products, lardaceous liver, cirrhosed kidney, vaccine matter, pyæmic abscess of the spleen; and in a large proportion of the cases produced tuberculosis of the various organs. Sanderson and Fox also produced tuberculosis by inserting setons of cotton-thread under the skin of these animals without inoculating them with any morbid material. Waldenburg, Cohnheim, and Fraenkel found that in the Pathological Institute at Berlin all the Guinea-pigs into whose abdominial cavities they introduced pieces of cork, paper, and cotton-thread, etc., became tuberculous, and concluded that the formation of a suppurative inflammatory focus is sufficient to render certain animals tuberculous, and therefore the non-specific character of tubercle. But a repetition of these experiments, with antiseptic precautions, at a later date, led Cohnheim and Fraenkel to modify their opinions. Schottelins, of Würzburg, produced granular pulmonary tuberculosis in dogs, by making them respire air charged with pulverized phthisical sputum. But he produced similar results with air charged with the

* Williams, "Pulmonary Consumption," etc., London, England. P. Blakiston, Son & Co., Philadelphia, 1887.

expectoration of bronchitis, Limburg cheese, and with vermilion.

Rindfleisch * calls attention to the remarkable similarity between the predisposition of "certain animals" for tuberculosis, and the occurrence of tuberculosis in a certain group of persons—the scrofulous. He declared that any large-celled infiltration of a tissue is to be regarded as tuberculous or scrofulous in character.

Ziegler † demonstrated in 1875 that neither giant-cells nor epithcloid cells are exclusively confined to tubercle, but are to be found in all granulations.

"In 1874 Demet ‡ and Paraskova Zablonus, of Syra, in Greece, succeeded in inoculating a man of fifty-five with tuberculosis. The patient was dying of gangrene of the left foot through obliteration of the femoral artery. Phthisical sputum was inserted into the upper part of the right leg, the lungs having been previously examined and pronounced perfectly healthy. Three weeks after the inoculation, signs of commencing induration of the right apex were detected, and seventeen days later (i. e., thirty-eight days after inoculation) the patient died of gangrene. The autopsy showed seventeen tubercles, varying in size from a mustard-seed upward, at the right apex, and a smaller number at the left apex, all evidently of recent formation. . . . Klebs, firmly convinced of the specific nature of the tubercle, had described an actively

---

* Rindfleisch, in Ziemssen's "Cyclopædia," etc.

† Ernst Ziegler, "Ueber die Herkunft der Tuberkelelemente," etc., Würzburg, 1875.

‡ Williams, "Pulmonary Consumption," etc., Philadelphia, 1887.

moving organism as its cause. . . . Schüller and Toussaint had pictured a spherical micrococcus in connection with the disease. . . . Aufrecht had found more than one form of organism, and thus paved the way for the next step, which was the discovery of Robert Koch."

Koch,* in 1882, judging from the results which had been recently attained concerning the etiology of many infective diseases, considered it not unlikely that the cause of tuberculosis might also be found in some micro-organism. In his investigations into the etiology of tubercle he followed the method by which the parasitic nature of splenic fever was so effectually established. He first turned his attention toward proving the presence of a pathogenic organism, thence passing on to isolation and inoculation experiments.

"Koch,† having by means of certain aniline dyes detected the bacillus tuberculosis, succeeded through a series of ingenious cultivations in procuring it pure and simple. He first took tubercle, and, after washing it with a solution of corrosive sublimate, removed the outer layers and separated a portion, into which he might fairly expect that no bacteria of putrefaction had penetrated. This he spread over a nutrient soil, consisting of the blood-plasma of the ox, which had been previously sterilized by boiling in a test-tube. The coagulum of this, with the tubercle added, was introduced into a test-tube with a cotton-

* "The Etiology of Tuberculosis," by Dr. Robert Koch, vol. ii, Berlin, 1884, translated by Stanley Boyd, F. R. C. S., New Sydenham Society, London, England, 1886.

† Williams, "Pulmonary Consumption," etc.

wool plug, and kept in an oven at a temperature of 98·6 to 100·4° Fahr. Nothing appeared during the period of incubation of the ordinary bacteria of putrefaction, but at the end of ten days there were seen on the dry surface of the coagulum a number of very small points or dry-looking scales surrounding the pieces of tubercle, spread out in circuits more or less wide, according to the distribution of the tubercle-fragments. After a few weeks' more exposure these crusts ceased to enlarge, and were then transferred to a fresh test-tube containing blood-plasma similarly prepared. After another interval of ten days the scales appeared, became confluent, covering more or less of the surface of the coagulum, as the seed was scattered, and so from test-tube to test-tube the experiment was carried out, under the most vigorous antiseptic conditions, as many as a dozen times, and for a period extending over one hundred and fifty days. With the results of these culture experiments two hundred rabbits and Guinea-pigs were inoculated, the places selected being under the skin, the peritoneal cavity, or the anterior chamber of the eye. With one exception all these animals acquired tuberculosis of the lungs, liver, spleen, and other organs, the tubercles having the structure of true tubercle, and including giant-cells, which latter were found to contain bacilli."

According to Powell,* "the main facts with regard to

* "Diseases of the Lungs and Pleura," by R. Douglas Powell, M. D., London, England. William Wood & Co., New York, 1886.

the life-history and potentialities of the tubercle-bacillus may be stated as follows:"

"1. The tubercle-bacillus is a minute rod-shaped fungus, measuring from 0·003 to 0·0035 millimetre in length, and about one third that measurement in thickness. The rods are straight and slightly curved, with rounded ends, and often inclose bright, spherical, spore-like granules of uniform size, arranged in linear series, and separated from one another by hyaline intervals. After having been stained with methyl blue, fuchsin, or magenta, and then washed in nitric acid, ten per cent, they retain the original dye, and are thus distinguished from putrefactive or other bacilli.

"2. This organism is only capable of growth and multiplication under culture in blood-serum or animal broth, at a constant temperature of 30° centigrade. It is of comparatively (to other bacteria) slow growth, and is unable to continue its development in decomposing fluids in the presence of more rapidly growing bacteria (Koch).

"3. All the conditions essential for the development of the bacillus are, so far as its life-history is known, alone to be found naturally in the animal body.

"4. The bacillus is, however, of very tenacious vitality, and will preserve its virulence and capacity for development for six weeks or longer in decomposing sputum, for six months or longer in the dry state.

"5. If a minute portion of bacillus containing matter be placed upon a neutral culture surface, and allowed to germinate, and if a fragment of the product of germination be similarly cultivated on a fresh surface, and so on for many generations, all foreign germs being excluded, the last product, if inoculated into an animal, will be as potent in producing tuberculosis as the first.

"6. The bacilli, whether derived from free cultivation or from tubercle, if intimately diffused in water, and scattered in the form of spray through an atmosphere in which animals are placed so that they inhale it, will produce tuberculosis in them.

"7. In the sputa of all cases of well-marked phthisis the bacilli are to be found.

"8. In cavities in the lungs of tubercular or caseous pneumonia—i. e., of phthisical origin—whether large or minute, bacilli are invariably to be found.

"9. In caseous and catarrhal pneumonic consolidations of the lung, excepting in the immediate neighborhood of cavities, large or minute, bacilli are sparse and rather difficult to find; large fields of sections may be traversed without discovering them; yet this material is virulent in producing tubercle when inoculated.

"10. In the granulations of miliary tuberculosis bacilli are very generally but not invariably to be found, and often only in small numbers. In their most recent researches upon the artificial inoculation of Guinea-pigs with tuberculosis (bacilli-containing) sputum, Drs. Klein and Gibbes found that the tubercular lesions contained but few and in many instances no bacilli.

"11. The result of inoculations made with dry bacillus-culture by Koch and many others, with the most minute precautions, have with much reason been accepted as proving the organism to be *per se* the virus of tubercle."

Jaccoud * (1880) admits two distinct varieties of pulmonary phthisis: one the inflammatory, or pneumonic form; the other the chronic, or ordinary form of the complaint. He states that— .

"1. Caseation is at all ages the result of debility.

"2. The origin of true tubercle is the result of debility.

"3. The common forms of accidental irritation of every kind, affecting the larynx, bronchial tubes, or lungs, have a deleterious effect upon tuberculosis and phthisical lesions. This may happen in three ways:

"Firstly, in those who are healthy, but in whom predisposition exists, such irritation favors the development of tubercles, or of the inflammatory changes which produce phthisis.

"Secondly, in those already affected it gives rise to a fresh development of tubercles.

"Thirdly, it aggravates and hastens the course of preexisting disorders.

"4. Fever is a process of consumption."

Loomis † (1884) states that "the essential pathological change of chronic phthisis is consolidation and induration of lung-substance. Tubercles may or may not be its primary lesion, and when present they may constitute but a small part of the morbid processes." Re-

---

* "Curability and Treatment of Pulmonary Phthisis," by S. Jaccoud, translated by Montague Lubbock, London, England. D. Appleton & Co., New York, 1885.

† Loomis, "Practical Medicine," etc. William Wood & Co., New York, 1884.

ferring to the part played by the tubercle-bacillus, he says: "The case at present may be stated as follows: The presence of a distinct bacillus in connection with tubercle, and its absence in all other morbid conditions, are generally confirmed by the most competent observers. The etiological relation of this bacillus to phthisis still rests solely upon Koch's demonstration."

Flint* (1885) defines two forms of chronic phthisis—tubercular and fibroid. He states that "heretofore pulmonary phthisis and pulmonary tuberculosis were considered as convertible terms, but, adopting Virchow's theory, in a certain proportion of cases pulmonary phthisis is not a tuberculous disease. Hence arose a variety of names denoting non-tuberculous phthisis, such as chronic broncho-pneumonia, chronic lobular pneumonia, catarrhal pneumonia, cheesy pneumonia, etc. These names have shared the fate of the theory from which they originated, the latter, at the present time, having but few supporters in any country. It is convenient to distinguish the morbid product which is characteristic of pulmonary phthisis as a tuberculous product, and it will be so distinguished in this article."

Flint states that "clinical experience fails to furnish positive proof of the communicability of phthisis."

Powell † (1886), referring to the etiological relation of the tubercle-bacillus to phthisis, says: "Notwithstanding the apparently insurmountable antagonism between those

---

* Flint on Pulmonary Phthisis, in Pepper's "System of Medicine," vol. iii. Lea Bros. & Co., Philadelphia, 1885.

† Powell, "Diseases of the Lungs," etc.

who adhere to the essentially bacillus nature of phthisis, and those who do not, there is a neutral ground where the two views meet, and where they may perhaps ultimately agree. Even Koch himself believes that certain pathological changes are, if not necessary, at least highly favorable to the reception of the germ."

II.

GEOGRAPHICAL DISTRIBUTION.

ACCORDING to Hirsch,* "Corresponding to the prevalence of consumption at all times is the universality of its geographical distribution at present. It extends over every part of the habitable globe. It may be designated an ubiquitous disease in the strictest meaning of the term. . . . Taking the mean death-rate of the whole of a population to be 22 per 1,000, and the average of deaths from phthisis of the lungs to be 3 per 1,000, we find that the deaths from consumption are nearly one seventh of the whole mortality (or in the ratio of 3 to 22)." Estimating the total yearly mortality of the world to be 35,000,000, we find that about 5,000,000 deaths are due to consumption, being the greatest number reported by reliable observers as due to any single cause of death.

The following tables of death-rates from pulmonary phthisis, and other data regarding its geographical distribution in countries other than the United States; have been taken from Hirsch's great work:†

---

* Hirsch's "Hand-book of Geographical and Historical Pathology." New Sydenham Society's translation, London, England, 1886.

† Ibid.

Table of Death-Rates from Pulmonary Consumption.

LOCALITY.	Period.	Deaths per 1,000 inhabitants.
Norway	1871–'75	2·5
Christiania	1866–'75	3·4
Sweden	1861–'76	3·5
Stockholm	"	4·1
Falun	1861–'65	3·0
Denmark.		
Copenhagen	1876–'83	3·0
Five largest towns	"	2·6
Twenty-four medium towns	"	2·2
Twenty-five smallest towns	"	2·1
Germany.		
Northeast coast and German plain.		
Königsberg	1877–'80	2·8
Dantsic	"	2·5
Stettin	"	2·6
Lübeck	"	2·6
Kiel	"	2·9
Posen	"	3·0
Breslau	1869–'78	3·7
Frankfort-on-Oder	1877–'80	3·5
Berlin	1869–'82	3·8
Magdeburg	1877–'80	3·8
Halle	"	2·7
Leipsic	"	3·5
Northwest coast and German plain.		
Hamburg	1871–'83	3·4
Altona	1877–'80	3·8
Bremen	"	3·2
Brunswick	1864–'73	4·0
Hanover	{ 1877–'80	4·3
	"	3·8
Central and southern hill-country.		
Dresden	"	3·8
Chemnitz	1870–'80	2·9
Erfurt	1877–'80	2·3
Gotha	"	2·5
Cassel	"	3·7
Würzburg	1871–'79	5·2
Nuremberg	1877–'80	4·7
Augsburg	"	3·9
Munich	"	4·0
Stuttgart	1873–'82	2·8
Plain of upper Rhine.		
Frankfort	1863–'83	3·5
Wiesbaden	1877–'80	4·0
Mainz	"	3·9
Darmstadt	"	3·7

Table of Death-Rates from Pulmonary Consumption—(continued).

LOCALITY.	Period.	Deaths per 1,000 inhabitants.
Mannheim	1877-'80	4·0
Carlsruhe	"	3·8
Strassburg	"	3·5
Metz	"	3·5
Plain of lower Rhine.		
Dortmund	"	4·7
Bochum	"	5·7
Hagen	"	6·3
Crefeld	"	5·8
Düsseldorff	"	3·5
Elberfeld	"	4·0
Barmen	"	4·5
Remscheid	"	8·8
Gladbach	"	7·3
Cologne	"	4·4
Bonn	1867–'72	3·5
Coblenz	1877–'80	4 3
Aix-la-Chapelle	"	3·8
Treves	"	4·7
Austria.		
Prague	1865–'74	8·5 (?)
Brünn	1873–'74	9·9 (?)
Linz	"	8·9 (?)
Vienna	1865–'74	7·7
Trieste	1870–'74	4·5
Pesth	1872–'75	6·9
England	1872–'73	2·2
London	1859–'69	3·2
Southeastern counties	"	2·6
Southern inland counties	"	2·3
Eastern counties	"	2·4
Southwestern counties	"	2·2
Western inland counties	"	2·2
Northern inland counties	"	2·4
Northwestern counties	"	3·2
Yorkshire	"	2·8
Northern counties	"	2·7
Wales	"	3·1
Scotland.		
Edinburgh	1857–'61	3·0
Leith	"	2·0
Glasgow	"	4·0
Dundee	"	3·4
Belgium	1856–'59	4·1
Brussels	1864–'78	5·6
Antwerp	1868–'74	3·3
Liége	1865–'74	4·0

Table of Death-Rates from Pulmonary Consumption—(continued).

LOCALITY.	Period.	Deaths per 1,000 inhabitants.
Holland	1869–'74	2·4
North Brabant	"	2·4
Herzogenbusch	"	2·8
Breda	"	3·3
Gelders	"	2·4
Arnhem	"	2·8
Nymwegen	"	2·4
South Holland	"	2·2
Gravenhaag	"	2·4
Delft	"	2·7
Leyden	"	2·6
Rotterdam	"	2·7
Gonda	"	2·2
Dordrecht	"	2·2
North Holland	"	2·3
Amsterdam	"	2·5
Alkmaar	"	2·9
Haarlem	"	3·0
Seeland	"	1·8
Middleburg	"	2·4
Utrecht	"	2·6
Utrecht	"	3·2
Friesland	"	2·5
Leeuwarden	"	2·7
Overyssel	"	3·2
Zwolle	"	3·3
Deventer	"	3·6
Groningen	"	2·3
Groningen	"	2·7
Drenthe	"	3·0
Limburg	"	2·3
Maestricht	"	2·9
Switzerland	1865–'69	1·8
Zürich	"	2·4
Winterthur	"	2·5
Chur	"	3·0
Bern	"	3·9
Geneva	"	2·2
France		
Paris	1872–'77	4·2
Italy		
Venice	1862–'85	4·0
Padua	1872–'77	2·8
Milan	1875–'78	3·8
Turin	1869–'76	2·7
Genoa	1875–'78	2·0
Verona	1874–'78	2·0

Table of Death-Rates from Pulmonary Consumption—(continued).

LOCALITY.	Period.	Deaths per 1,000 inhabitants.
Rome	1874–'78	3·4
Bologna	1875–'78	3·8
Naples	"	2·7
Palermo	1873–'78	2·6
Messina	1876–'78	3·0
Catania	1877–'78	1·4
Malta	1822–'34	3·3
East India.		
Native troops	1850–'60	3·0
Australia.		
Melbourne	1865–'70	2·2
St. Helena	6 years.	2·2
Algiers	1852–'59	2·9
Brazil.		
Pernambuco	....	5·2
Rio de Janeiro	1855–'58	5·0
Desterro.		
S. Catarina	1862	3·9
Uruguay.		
Montevideo	1871–'74–'75	4·0

Although these statistical data should not be credited, as we have seen, with more than limited value, yet they supply incontrovertible evidence that there are differences in the frequency of consumption, sometimes even very considerable differences, in the various divisions of a country, and even at various points within a small area; and that is the conclusion which is fully borne out by information come by in other ways as to the amount of the disease in different parts of the world.

A very remarkable fact in the geographical distribution of consumption on *European soil* is its rarity in many of the islands and coast districts within the more northern latitudes, such as *Iceland*, the *Faröe Islands*, the *Hebrides*, the *Shetland Islands*, and places in *Norway*.

Schleisner, writing of Iceland, says: "According to the unanimous testimony of practitioners in the island, consumption does indeed occur there, although remarkably seldom. In my own practice I have most carefully examined every patient who complained of even the slightest trouble in the chest, and, out of 327 persons suffering from chronic diseases of the organs of respiration, I found only three with phthisis, one of these being a person of Danish extraction." That statement is borne out in the more recent writings on the state of health in Iceland by Leared, Hjaltelin, and Finsen. It would appear that it is not with any national peculiarity that we have here to do, from the fact that Icelanders who migrate to Denmark fall into consumption not unfrequently. To the same effect is the information given by Manicus and Panum for the *Faröe Islands;* among 100 patients examined by the latter, there were only two with phthisis. In the *Hebrides* this disease is almost unknown, in the *Shetland Islands* it is said to have been not at all common until recent years. Consumption appears to be more common in *Sweden* than in Norway. As a general rule, consumption is more prevalent in the southern than in the northern regions of Sweden, although the difference is not so great as in Norway.

In the islands and mainland of *Denmark* the disease stands at about the mean average of frequency, according to the results of Lehmann's inquiries. The same appears to hold good for the northern governments of *Russia in Europe,* although the very meager and somewhat vague information from that country does not enable us to come to any conclusion with certainty. In St. Petersburg the disease is not more frequent, at all events, than in the average of large European cities; in Finland and the Baltic provinces there is little of it, except in the large towns; from the central and southern parts of the country we hear of it as frequent in Novgorod, Viatka, Kasan, Kursk, Kischniew, Odessa, Sebastopol, and Astrakhan; in Orenburg, also, it is not altogether rare.

But among the Kirghiz of the steppes it is quite unknown; so much so, that Neftel did not see a single case of phthisis among them during a period of several years. In the Caucasus, consumption is prevalent mostly in the higher parts of the interior; it is but rarely seen along the course of the Rion or on the Black Sea coast.

The following table shows the distribution of phthisis in *North Germany:*

Mortality from Phthisis in Prussia from 1875 to 1879.

DEPARTMENT.	Inhabitants per square kilometre.	Deaths from phthisis per 1,000 inhabitants.	Ratio in the urban population.	Ratio in the country population.
Baltic.				
Gumbinnen	47	1·96	2·77	1·84
Königsberg	52	1·74	2·49	1·45
Danzig	68	1·74	2·39	1·41
Marienwerder	45	1·61	2·54	1·35
Stettin	57	2·39	2·90	2·08
Köslin	39	1·85	2·58	1·60
Stralsund	52	2·57	3·21	2·12
Schleswig	58	3·22	3·31	3·18

Mortality from Phthisis in Prussia from 1875 to 1879—(continued).

DEPARTMENT.	Inhabitants per square kilometre.	Deaths from phthisis per 1,000 inhabitants.	Ratio in the urban population.	Ratio in the country population.
Warthe and Oder.				
Posen	59	2·30	2·96	2·04
Bromberg	50	2·20	3·13	1·85
Breslau	109	3·07	3·73	2·75
Liegnitz	73	2·52	2·98	2·35
Oppeln	104	2·55	2·99	2·45
Frankfurt	55	2·54	3·08	2·25
Prussian Saxony, the Mark, etc.				
Potsdam	53	2·53	2·88	2·33
Magdeburg	76	2·79	2·98	2·65
Merseburg	88	2·29	2·63	2·16
Erfurt	109	2·70	2·69	2·70
North Sea.				
Hanover	74	3·99	3·38	4·44
Hildesheim	80	3·02	2·66	3·21
Luneburg	33	3·47	3·85	3·39
Stade	47	4·01	3·18	4·20
Osnabrück	44	5·14	4·87	5·22
Aurich	64	3·67	3·31	3·79
Lower Rhine.				
Cologne	164	5·11	4·76	5·34
Treves	85	3·55	3·53	3·56
Aix	121	4·02	3·64	4·59
Coblenz	92	4·33	4·26	4·35
Düsseldorf	267	5·29	5·22	5·29
Münster	61	5·17	6·50	4·70
Minden	91	4·71	4·73	4·90
Arnsberg	127	4·86	5·46	4·51
Upper Rhine.				
Cassel	78	3·17	3·48	3·03
Wiesbaden	122	3·98	3·82	4·08

It follows from this table, which is compiled from the painstaking work of Schlockow, that the disease is much less common in the territory of the Vistula, Oder, and Elbe, where the mortality is from 1·61 to 3·22 per 1,000, than in the territory of the Weser and Rhine, where the mortality is from 2·66 to 6·50 per 1,000. In the kingdom of Saxony, the yearly mortality, according to the figures for the years 1873–'80, is 2·40 per 1,000, or nearly as much as in the Prussian department of Erfurt, the maximum of 2·81 to 2·77 falling in the circles of Dresden and Leipsic, with a minimum of 1·75 in the circle of Bautzen, and one of 0·90 per 1,000 in the circle of Zwickau.

In the kingdom of Bavaria, according to the statistical returns of 1867–'75, the annual mortality from phthisis has a mean of 3·14 per 1,000; but the results can not be compared with those obtained from other parts of Germany, inasmuch as they relate to all deaths from pulmonary consumption, from general tuberculosis, and from wasting in persons over fifteen, which should be compared separately to be correct. The Bavarian departments with the largest death-rates are Lower Franconia, with 3·61; Middle Franconia, with 3·49; and Upper Franconia, with 3·24. Next comes the Palatinate, with 3·20; the Upper Palatinate, with 3·12; Swabia and Bavaria, each with 3·06; while the minimum of 2·40 is reached in Lower Bavaria.

In the grand duchy of Baden, according to the figures for the years 1874–'81, the mean annual death-rate from consumption is 2·78 per 1,000 inhabitants, being distributed in the various circles as follows:

Mannheim	3·87	Constance	2·65
Carlsruhe	3·41	Lorrach	2·54
Baden	3·28	Mosbach	2·37
Freiberg	3·05	Villingen	2·36
Heidelberg	3·04	Waldshut	2·24
Offenburg	2·89		

In the grand duchy of Hesse the mean annual death-rate from phthisis was 2·73 per 1,000 (according to the figures of 1877–'81), Starkenburg having 2·83, Upper Hesse 2·42, and Rhenish Hesse 2·82. We are obliged to notice the very remarkable comparative immunity of the elevated regions of Germany from consumption as contrasted with the low plains. The same relative infrequency of the disease is met with in the mountainous parts of Austria; whereas in the level country and among the lower valleys of Galicia, Upper Austria, Styria, and Carinthia, consumption is very prevalent.

In *England*, as will be seen from the table, the heaviest mortality (2·8 to 3·5 per 1,000) falls in London and in certain of the northern and northwestern counties (Notts, Derby, Cheshire, Lancashire, West Riding, Durham, Northumberland, Cumberland, South Wales, and North Wales). The smallest mortality (1·8 to 2·2 per 1,000) is found in the southwestern and inland southern counties of Wilts, Dorset, Somerset, Herts, and Bucks; in the western inland counties of Gloucester, Hereford, Shropshire, Staffordshire, and Worcester; in the counties of Rutland and Lincoln, the North Riding, and the mountainous district of Westmoreland.

From *Scotland* we have accounts of the very rare occurrence of the disease in the western Highlands.

Of *Ireland*, Wylde says that phthisis is "by far the most fatal affection to which the inhabitants of this country are subject."

In *Holland*, as the figures in the table show, the

chief centers of phthisis are in the northeastern provinces of Overyssel and Drenthe, next in order being Utrecht, Friesland, and North Brabant; it is rarest in Zeeland and next rarest in South Holland and Limburg. Among the larger towns, those most afflicted with consumption are Deventer, Zwolle, Breda, Utrecht, Haarlem, and Maestricht.

In *Belgium*, as we learn from Meynne, there is most of it in the industrial centers, such as Brussels, Ghent, Bruges, Liége, St. Nicolas, Verviers, and Ypres; while we have more special details of its extensive prevalence in Antwerp, Boom, Contich, Mechlin, Haeght (Brabant), Ecloo (West Flanders), Courtray, Furnes and Dikmude (both in East Flanders), and in Beauraing (Namur).

Few countries of Europe enjoy on the whole so favorable conditions as *Switzerland* in respect of the infrequency of consumption. According to Müller's inquiries, the total mortality from phthisis is 1·86 per 1,000 inhabitants. It was distributed as follows in the several cantons:

CANTON.	Years observed.	Deaths per 1,000 inhabitants.
Baselstadt	3	3·57
Grisons	2 to 5	2·50
Geneva	3 to 5	2·40
Neuchâtel	2 to 5	2·40
Schwyz	3 to 5	2·30
Schaffhausen	5	2·10
Aargau	4 to 5	2·00
St. Gall	4 to 5	2·00
Zürich	5	1·96
Bern	3 to 5	1·90
Ticino	3 to 5	1·90

CANTON.	Years observed.	Deaths per 1,000 inhabitants.
Vaud	2 to 5	1·70
Zug	5	1·60
Basel Land	5	1·50
Thurgau	1	1·45
Uri	3	1·40
Unterwalden	3 to 4	1·40
Appenzell	4	1·35
Vallais	5	1·20
Freiburg	2	0·81

The comparatively high mortality of Baselstadt, Geneva, and Neuchâtel, is explained by the industrial character of the respective towns; that of the Grisons by the common occurrence of the disease in Chur (3 per 1,000); that of Canton Schwyz, as well as the small figure for Canton Freiburg, most probably by errors in the returns.

The information as to the extent of phthisis in *France* comes in fragments from the various parts of the country; and we are unable to form a general estimate of it. As in other countries, it is the great centers of commerce, trade, and manufacture that form the chief seats of the malady. Whether, as Lambard alleges, "it is more common in the northern and western departments than in the southern, eastern, and central," *it is difficult to decide.*

According to Drysdale,* "out of 1,000 deaths occurring in Paris weekly, 200 were from phthisis, and 25 were from external tuberculosis. Dr. Thouvenin has

* "Climatic Treatment of Consumption," by Dr. C. R. Drysdale, London, England. British Medical Society, August, 1887.

shown that 65 per 1,000 of the deaths among the rich were caused by it, while 230 per 1,000 among the poor succumbed to the disease."

For *Spain* and *Portugal*, also, there are only fragmentary notices of the distribution of consumption. It is prevalent to a very great extent on the central plateau of Spain (in New Castile and Estremadura) as well as in certain of the larger ports on the western and southern coasts, such as Barcelona, Valencia, Cadiz, and Gibraltar, Hennen's statement that phthisis is "truly endemic" in these having been subsequently confirmed by Chervin. To the same effect is the opinion of Wallace, of Trogher, and of Brant, on the frequency of the malady in Lisbon, the last-named remarking that it takes a foremost place among the diseases of the people in the plains of Portugal and in the densely populated towns.

The following are the official figures of the mortality from consumption in *Italy*, for the years 1881–'83, according to provinces:

Table of the Death-Rate from Phthisis in Italy per 1,000 Inhabitants.

Italy (general average) .... 2·45

Lombardy	3·34	Marches	2·06
Latium (Rome)	3·18	Umbria	1·87
Toscana	3·16	Sardinia	1·78
Piedmont	2·86	Apulia	1·63
Æmilia	2·75	Sicily	1·48
Liguria	2·71	Abruzzi (Molise)	1·42
Campania	2·43	Calabria	1·36
Venetia	2·28	Basilicata	0·89

Assuming that these figures are trustworthy, or that the same unavoidable sources of error have recurred in all the localities, it follows that there is a considerable preponderance of the disease in the northern provinces, and that its frequency diminishes very decidedly as we go southward.

In *Roumania* consumption is unusually common, according to the testimony of all authorities. In *Turkey*, also, and particularly in Constantinople, the disease is far from rare, being met with very extensively among the Turkish troops. In the island of *Cyprus* it is said to be almost unknown. In like manner, *Greece* would seem to be comparatively well off in respect to the rarity of consumption; thus, it is spoken of as being seldom met with in Laconia. According to casual notices from the countries of *nearer Asia*, such as the plateau of *Armenia*, the coast-plains of *Syria*, and the tableland of *Persia*, phthisis would seem to be a comparatively rare trouble.

As regards Armenia, Wagner says that the disease is seen only in persons who have migrated from Arabia, Mesopotamia, or the countries of the negro. Polack says the same of the Persian plateau, where the indigenous inhabitants enjoy an almost absolute immunity. The rarity of the disease on the plains of Syria is attested by Yates, Robertson, Tobler (for Jerusalem), and Barret (for Beyrout). It is more frequently seen in the Lebanon, in the neighborhood of Baalbec and of Aleppo, where Guys tells us that it is truly endemic; also on the Arabian shore of the Red Sea, especially among those Bedouins "who have exchanged the tent," as Pruner says, "for the stone-built house."

As in the countries just spoken of, so also in *India*, the prevalence of phthisis can not be given in figures. It is, on the whole, rarer in that part of the world than in the temperate zone of the Eastern Hemisphere, but by no means so rare as the earlier observers supposed, from their imperfect means of diagnosis.

In some districts it is, in fact, common, particularly among the English troops; as in the plain of Upper Bengal stretching along the foot of the Himalaya, in the district of Madras (among natives as well), at places on the Malabar coast such as Cochin and Cannanore, at Bombay, in certain localities of the Northwest Provinces, and in the Punjaub, where Hinder found it very prevalent among the natives of Amritsur. It does not occur so often in Lower Bengal and Assam nor in Upper Sinde, extremely seldom on the plateaus of the Western Ghauts at levels of 4,000 to 7,000 feet, or in the Nilghiri Hills, or on the northern and southern slopes of the Himalaya.

On one point all the authorities in India are agreed—that the disease in that country is of an extremely pernicious type; and the same is true of it in all other tropical regions of Eastern Asia, including *Ceylon*, the *East Indies*, *Farther India* (*Cochin-China* in particular), *China*, and *Japan*. In Ceylon, consumption is found mostly among the black population. In the Malay Archipelago the disease is far from rare, as we may infer from the above reference to the great mortality caused by it among the native troops; all the authorities write in that sense, mention being made more particularly of its common occurrence in Amboina and in the Philippines, especially at Manilla.

In Sumatra it would seem to be rare. The French medical officers are entirely agreed as to the great frequency and pernicious type of phthisis in Cochin-China.

There are reports to the same effect from various parts of China, such as Canton and Hong-Kong, Chang-fu, Tientsin, and Pekin; at other ports, such as Shanghai and Hankow, it is not so common.

In Japan, phthisis holds one of the first places among the causes of death.

Consumption is prevalent to a most disastrous extent among the native races of the *Southern Pacific.* We have more particular accounts for Fiji and Tonga, Samoa, Tahiti, the Marquesas, and Hawaii (Honolulu). In New Caledonia the death-rate from consumption among the Kanakas is estimated at two fifths of the mortality from all causes. Almost all the authorities are of opinion that the great prevalence of the malady in these islands dates from the time when the natives began to come into more intimate relation with European immigrants, and therewith to make considerable changes in their mode of life; and that opinion is borne out by the fact that in the Hawaiian Islands, where phthisis at the present time commits great ravages among the natives, it was of rare occurrence forty or fifty years ago. On the other hand, it follows, from Wilson's account (1806) of the state of health in Tahiti, that phthisis had been widely prevalent in that group as early as the beginning of the century;

and there are accounts to the same effect from the Tonga group, New Caledonia, and other of the archipelagoes of Polynesia.

The reputation that *Australia* used to enjoy for the rarity of consumption, and for the favorable influence of its climate upon the course of the malady, has of late been shown to be a mistaken one. In Victoria, where the disease, it is true, has become a good deal more common only within recent years, the mortality from phthisis in 1866 was six per cent of the mortality from all causes, while in Melbourne itself the death-rate rose between 1865 and 1869 from 2·22 to 2·52 per 1,000 of the population. In Tasmania it would not seem to be common; during five years' practice Hall saw 235 cases at Hobart, only 37 of whom had been born in the colony, all the rest being immigrants from Europe. In New Zealand phthisis has made frightful ravages among the Maoris, and has been one of the chief causes of the gradual extinction of that race.

Among the islands on the eastern side of Africa, *Mauritius* and *Rèunion* are the two most subject to consumption. On Nossi-Bé, also, the malady is not uncommon among the colored races, particularly the Caffris. In *Madagascar* and *Mayotte* it is as common as in Europe, and rapidly fatal, as it mostly is in the tropics. In Zanzibar, Lostalot did not happen to see many cases, but it is said to be especially common among the Arabian women of the higher class. In *Cape Colony* phthisis is oftenest met with among the Hottentots inhabiting the plains nearest the coast; in other classes of the population it is much rarer than in the East African islands within the tropics, just spoken of; while on the interior plateau of Southern Africa it hardly occurs at all.

There is a lack of information of a trustworthy kind as to the state of health on the southern part of the *west coast of Africa*, the coast of Lower Guinea. Around the *Bights of Benin and Biafra* (country of the Cameroons and of the Gaboon), as well as in the adjoining island of *St. Thomas*, it appears, from the entirely trustworthy writings of Daniell, that phthisis is widely prevalent and very malignant among the negroes. As regards the French settlements on the Gaboon coast that statement is fully borne out by the French medical practitioners; and we have an account to the same effect regarding its occurrence on the island of *Fernando Po*.

The circumstances in respect to phthisis are more favorable among the natives of the *Gold Coast* and of *Sierra Leone*. Among the islands of the *Cape Verde* group, Mayo is but little afflicted by the malady, while St. Jago and Fogo are much subject to it. We are unable to form a trustworthy conclusion of the extent of its prevalence on the *coast of Senegambia;* the older accounts, by Thevenot and Berville, speak of it as very rare; and so far in agreement therewith Chassaniol, Borius, Gauthier, and others state that it is at all events rarer than in France. On the other hand, Carbonell shows, from the statistics of mortality at St. Louis, that consumption makes up a considerable proportion of the deaths; while Defaut, writing of Gorée, says: "La phthisie pulmonaire est frequent, et on peut dire que toutes les classes d'habitants y sont exposées; en effet, les noirs fournissent un contingent considérable, et les blancs en sont souvent atteints."

Among the nomade tribes of the interior, as Carbonell states, the malady is of rare occurrence. According to the same authority, the death-rate from phthisis among the French garrison, from 1862 to 1865, was 2·7 per 1,000 of the total strength, and from 1869 to 1871, 2·72; in a battalion of natives numbering 450 men there were 23 deaths from consumption in fifteen years, giving a mortality of 3·4 per 1,000. He adds that "la population noire du Sénégal est, comme partout ailleurs, très sujette à la phthisie." As in tropical countries generally, the disease among the Europeans resident there runs a very rapid course.

Several of the territories on the *north coast of Africa* enjoy a notable exemption from phthisis. In Mogador, Morocco, and other places on the coast of the sultanate of Morocco, the disease is mentioned as one that rarely occurs. In *Algiers*, also, it is a good deal less prevalent than in Europe, although the extremely favorable reports by French physicians of a former period as to its rarity in that country have not been fully confirmed. But all the more recent observers are agreed that phthisis is comparatively rare among the French, both civilians and military; while among the native population, and particularly among those living outside the towns, either occupied in agriculture or leading a nomadic life, it is still less frequently met with. The accounts from the province of Oran are especially favorable; in the capital town of that name, with a population of 25,000, not more than twelve cases of phthisis had occurred in eight years, and these were strangers, of whom only three or four had been taken ill subsequent to their arrival. In the last two years of the period not a single case had been known there. At two other places in the province, whose inhabitants numbered 2,130 Europeans and 4,300 natives, Gaucher met with only ten cases of phthisis in three years.

Pietra Santa, who writes of the province of Alger, and assigns to the capital the somewhat high average death-rate from phthisis of 2·7 per 1,000 inhabitants over a period of eight years, goes on to say of the agricultural or nomadic population of the province: "Tous les documents s'accordent à prouver, que la phthisie est extrémement rare chez les divers ébranchements de la race arabe." There is information to the same effect for the native population of the cultivated oases in the Algerian Sahara, particularly for the nomadic inhabitants of Great Kabylia, who enjoy, according to all authorities, an almost absolute immunity from consumption. In Algiers, as in many other tropical or sub-tropical countries, it is the negro race that seems to be most subject to the disease.

Along the seaboard of *Tunis*, *Egypt*, and *Abyssinia*, phthisis is found more often than in Algiers, among the natives as well as in others. In Egypt, as Pruner tells us, the disease becomes less in exact proportion as we proceed southward from the shore of the Mediterranean; in Central and Upper Egypt it is decidedly uncommon; but in Khartoum and Sennaar, as well as over the whole basin of the Nile beyond the tropic, it again becomes somewhat prevalent. The plateau of Abyssinia is almost free from consumption; Blanc states that he did not see a single case of it among thousands of patients during a lengthened residence in that country.

In the *Western Hemisphere* the inhabited regions within northern latitudes, and with an arctic climate, offer a marked contrast to the corresponding territories of Europe in respect to the great frequency of phthisis in them. In *North Greenland* that disease is one of the commonest causes of death. At a trading-station on the northern shore of *Hudson Bay*, phthisis is prevalent among the scanty population to an enormous extent, according to the evidence of a practitioner who had been five years on the station; and there are reports to the same effect from *New Archangel* and the *Aleutian Islands* (Alaska). It is common also in *Newfoundland*, *New Brunswick*, and *Canada*, in the last particularly among the native Indians (Stratton).

The elevated plains and mountain valleys of *Mexico* and other *Central American* countries, over 3,000 feet above the sea-level, enjoy, like those of the United States, an immunity from consumption more or less pronounced; on the other hand, the disease is very widely spread, and at the same time of a very malignant type, on their low plains and coasts.

Thus, while the authorities are unanimous in asserting the rarity of phthisis on the Anahuac or table-land of Mexico, and on the lofty plains of Honduras, San Salvador, Costa Rica, and Panama, we have particular information of its frequency and malignancy in Tampico and Vera Cruz (east coast of Mexico), in Campeché and Merida (coast of Yucatan), in Guaymas and Mazatlan (west coast of Mexico), on the Nicaragnan or Mosquito coast, and on the Panama coast.

The large amount of consumption in the *West Indies* had been remarked upon by many of the earlier observers, and their statements have been fully borne out by the more recent accounts from Cuba, St. Thomas, St. Martin, St. Vincent, St. Bartholomew, Guadeloupe, Martinique (where phthisis causes more deaths than any other disease except dysentery), St. Lucia, Barbadoes, and Trinidad. The disease is equally common and pernicious on the coast and plains of *Guiana*, whereas it is almost unknown among the natives inhabiting the mountainous part of the country.

Within the last fifty or sixty years there has been a very extensive diffusion of phthisis along the whole littoral of *Brazil*, from Pernambuco to Santa Catarina, as well as on the littoral of *Uruguay* and the *Argentine Republic*, and in the basin of the Rio de la Plata, including *Paraguay*. It is the large seaports that suffer most: thus, Beringer gives the mortality from phthisis in Pernambuco as 5·2 per 1,000 inhabitants, the negroes and mulattoes furnishing the greater part of it. In Rio de Janeiro, as we learn from Rey, the mortality from consumption has been on the increase; from 1855 to 1858, 14 per cent of the deaths from all causes (which came to 35 per 1,000 inhabitants) were from phthisis; whereas from 1867 to 1869, when the total death-rate had fallen to 24 per 1,000 inhabitants, that of phthisis had come to be 20 per cent of the total, or the comparatively excessive fraction of one fifth. In the interior, also, of Brazil, phthisis is very common, more especially in the larger towns. The mountainous region of the Argentine Republic and of *Bolivia*, like other mountainous regions, forms an exception. Even at Salta, Jujuy, and other places situated no higher than 1,000 to 1,200 metres (3,000 to 4,000 feet), the disease is rare; it is almost unknown in the capital towns of the provinces of Cochabamba, Chuquisaca, and Potosi, at elevations of 2,000 metres (6,500 feet) and upward, as well as throughout the whole mountain-range of Bolivia. In complete accord with the accounts of its frequency on the east coast of South America, and its considerable increase there within the last thirty or forty years, is the information about its progressive diffusion in the coast districts of *Chili*, *Peru*, and *Ecuador*. It is met with not uncommonly also among the deep valleys of the Andes, with a warm and moist climate, and in the forest region of Peru, even at elevations of 500 metres (1,600 feet); but the high plateaus are almost entirely free from it.

III.

GEOGRAPHICAL DISTRIBUTION OF CONSUMPTION IN THE UNITED STATES.*

LOCALITY IN RELATION TO DEATHS.

"In order to study the influence upon the health of the inhabitants exerted by peculiarities of topography, drainage, climate, etc., the country has been divided into regions, the physical characteristics of which are more or less distinct. This division was made by Mr. Gannett, the geographer of the census-office, and in the following account of these grand groups his descriptions of each of them are included, together with notes on peculiarities of climate, density of population, etc., and references to some of the causes of death which are either unusually frequent, or the reverse, for each locality." †

The first four of these regions, which comprise the whole Atlantic and Gulf coasts, possess primarily a sea-climate. In this region, to a greater or less extent, the extremes of heat and cold are lessened and mitigated by the presence of that great balance-wheel of temperature,

---

* Compiled from the tenth (1880) United States census and Government reports, and from Rand-McNally's "Atlas," Chicago, 1887.

† Tenth United States census.

the ocean. The atmosphere is moister and, as a rule, the rainfall is greater, than that of the country farther inland. This region, however, varies in its different parts very greatly in respect to temperature and surface, in such a manner as to produce very decided differences in its relations to certain causes of disease.

1. North Atlantic Coast Region.

This comprises a strip of land, from 50 to 75 miles wide, along the coast of Maine, New Hampshire, Massachusetts, Rhode Island, and Connecticut. The surface is mainly undulating and hilly, becoming less varied toward the south. The coast is bold and rocky in Maine, but mostly sandy and low in Massachusetts, Rhode Island, and Connecticut. There is comparatively little swamp or undrained land. The mean annual temperature is from 40° to 50° Fahr. The mean annual rainfall is from 40 to 50 inches. The mean elevation is from 100 to 500 feet, sloping toward the shore. The density of population is over 45 persons to the square mile, and over 90 in Massachusetts and Rhode Island. The colored population forms less than 7 per cent of the whole. The foreign population is from 20 to 35 per cent of the whole, except on the coast of Maine, where it is below 5 per cent.

The principal causes of death, which are reported as causing more than the average number of deaths out of the whole number reported, are, for this group: Scarlet fever, cholera infantum, old age, consumption, hydrocephalus, cancer, diseases of the nervous system, especially apoplexy and paralysis, diseases of the circulatory system, especially diseases of the heart, bronchitis, Bright's disease, drowning, and suicides.

The causes of death, in which the proportion is decidedly lower in this grand group than for the average of the United States, are: Measles, whooping-cough, enteric fever, diarrhœa and dysentery, malarial fevers, puerperal septicæmia, scrofula and tabes, dropsy, pneumonia, abortion, childbirth, diseases of the spleen, burns and scalds.

The proportion of deaths from consumption is high on the Maine coast, diminishing somewhat as we go south.

2. Middle Atlantic Coast Region.

This includes a strip of land comprising the coast counties of New York, New Jersey, Delaware, Maryland, and Virginia. The climate is somewhat milder than that of Grand Group I. The surface is low and sandy, and along the New Jersey coast we find characteristic sandy reefs, shoreward from which are lagoons, succeeded by extensive areas of swamp. Farther inland the country is low, nowhere rising more than 100 feet above the level of the sea. The mean annual temperature is from 45° to 50° Fahr. in the northern portion, and 55° to 60° in the southern portion. The mean annual rainfall is from 45 to 55 inches. The average density of population is over 45 to the square mile. In New York and Northern New Jersey it is over 90 to the square mile. In the northern part the colored population is below 7 per cent of the whole, while in the southern part it forms from 35 to 60 per cent of the population. The foreign population is below 5 per cent, except in New York and Northern New Jersey, where it is from 20 to 34 per cent.

The following are the causes to which are attributed a decidedly greater proportion of deaths out of all those reported than is the case for the average of the United States: Diarrhœa, cholera infantum, inanition, premature birth, stillbirth, debility, consumption, hydrocephalus, apoplexy, convulsions, diseases of the heart, bronchitis, dentition, Bright's disease, and peritonitis.

The proportion of deaths from the following-named causes is below the average: Measles, diphtheria, whooping-cough, enteric fever, malarial fever, erysipelas, dropsy, diseases of the brain, croup, pneumonia, abortion and childbirth, and accidents and injuries.

The proportion of deaths from consumption is comparatively high in Delaware, lower in New Jersey, New York, and Maryland, and lowest on the Virginia coast.

The proportion of deaths from different causes in this grand group is influenced to a great extent by the presence in it of the large cities of New York, Brooklyn, Baltimore, and Washington.

3. South Atlantic Coast Region.

This includes the coast counties of North Carolina, South Carolina, and Georgia, with extensive reefs inclosing large bays and sounds. A large proportion of the area is low and swampy. It includes that portion of the

States above mentioned which lies below what is called the "fall line"—that is, the line which forms the boundary of the metamorphic region. The mean annual temperature is from 60° to 65° Fahr.; the mean annual rainfall is from 50 to 60 inches. The density of population on the coast of North Carolina and the northern portion of South Carolina is from 6 to 18 to the square mile. For the rest of the group it varies from 18 to 45 per square mile. The average elevation above the sea is less than 100 feet. The proportion of foreign population is less than 5 per cent of the total population. The colored population is over 50 per cent. The following causes of death are given as causing a greater proportion of the whole number of deaths reported than the average for the United States, namely: Measles, whooping-cough, diarrhœa, worms (in the rural districts), dropsy, tetanus, trismus—nascentium, dentition, urinary calculus, childbirth, burns and scalds, exposure and neglect, and gunshot-wounds.

The following are the causes from which the proportion of deaths is reported as being less than the average, namely: Scarlet fever, cholera infantum, erysipelas, puerperal septicæmia, old age, consumption, hydrocephalus, cancer, diseases of the nervous system, diseases of the circulatory system, croup, bronchitis, pneumonia, Bright's disease, and other diseases of the kidneys, and diseases of the bones and joints.

The proportion of deaths from consumption is low on the Carolina coast, and slightly higher on the Georgia coast.

4. Gulf Coast Region.

This region includes the entire State of Florida and the coast counties of Alabama, Mississippi, Louisiana, and Texas. In Florida and Louisiana a large portion is uninhabited swamp-land. The mean annual temperature is from 70° to 75° Fahr.; the mean annual rainfall is over 55 inches. The density of the population is nowhere above 45 to the square mile, and over a large portion of this region it is below 6 to the square mile. The elevation above the sea is less than 100 feet, with the exception of a small part of the interior of Northern Florida, where it is from 100 to 500 feet. The colored population of this group forms from 35 to 60 per cent of the whole; the foreign population is below 5 per cent, except on the Texas coast, where it rises to 30 per cent and over.

The causes of death in this region, to which are attributed more than the average proportion of deaths, are chiefly: Whooping-cough, diarrhœa and dysentery, malarial fever, debility, consumption, dropsy, tetanus and trismus—nascentium, dentition, diseases of the liver, Bright's disease, childbirth, and gunshot-wounds.

The causes of death, to which in this region the proportion of deaths attributed is below the average, are as follows, namely: Measles, scarlet fever, diphtheria, enteric fever, cholera infantum, erysipelas, puerperal septicæmia, old age, croup, pneumonia and diseases of the respiratory system in general, and diseases of the bones and joints.

5. Northeastern Hills and Plateaus.

Grand Groups V, VI, and IX include the area of highlands stretching from northeast to southwest which has generally received the name of the Appalachian region. It comprises the broken, hilly country of Maine, the White Mountains of New Hampshire, and the Green Mountains of Vermont, the hills of central Massachusetts and of northern Connecticut, the Adirondacks and Catskills of New York, the multitudinous ridges and ranges of Pennsylvania, Virginia, West Virginia, the Carolinas, Tennessee, Kentucky, Georgia, and Alabama.

The northeastern Appalachian region, or Grand Group V, includes all that portion of Maine, New Hampshire, Massachusetts, and Connecticut not comprised in the coast strip, with all of Vermont, and the northern portion, including the Adirondacks, of New York. The area is by no means all strictly mountainous country. It includes a large amount of hilly, broken country. It was originally covered with dense forests, which have in the settled portions been largely cut away. The climate is severe, being affected comparatively little by the sea, and the mean annual temperature over most of this area is less than 45° Fahr. In some parts, although not the most thickly settled ones, it falls below 40° Fahr. The annual rainfall is from 35 to 45 inches. The mean density of population is below 45 per square mile. The elevation is mostly above 500 feet, and in considerable parts rises to mountains from 3,000 to 5,000 or even 6,000 feet in height. The colored population is below

7 per cent of the whole, and the foreign population is below 10 per cent.

The causes of death in this region, to which are attributed more than an average proportion of the deaths reported, are as follows: Diphtheria, old age, consumption, hydrocephalus, cancer, tumors, glycosuria, apoplexy and paralysis, diseases of the heart and Bright's disease, and diseases of the kidney and bladder.

The rate for consumption is very high in Maine, and is also high in Vermont, Massachusetts, and Connecticut; it is somewhat lower in New Hampshire and New York.

The causes of death in Grand Group V, to which are attributed less than an average proportion of the deaths reported, are mainly as follows: Measles, scarlet fever, whooping-cough, enteric fever, diarrhœa, dysentery, cholera infantum, malarial fever, puerperal septicæmia, premature birth, scrofula and tabes, convulsions, croup, bronchitis, dentition, childbirth, abortion, and accidents and injuries.

6. The Central Appalachian Region.

This comprises the Catskill region of southeastern New York, the central portion of Pennsylvania, and the western part of Maryland, and chiefly consists of narrow parallel ridges, with singularly uniform crests, broken by few gaps, and rising from 1,000 to 2,000 feet above the narrow valleys separating them, which, in their turn, are from 500 to 1,000 feet above the sea. The mean annual temperature is from 40° to 45° Fahr. The mean annual rainfall is from 35 to 40 inches. The density of the population is below 45 persons per square mile. The proportion of the colored population is below 7 per cent. The proportion of the foreign population is between 10 and 20 per cent.

The causes of death in this region, to which are attributed more than an average proportion of the deaths reported, are as follows: Scarlet fever, diphtheria, old age, cancer, diseases of the nervous system, more especially apoplexy, paralysis and convulsions, diseases of the heart, and railroad accidents.

The causes of death in this region, to which are attributed less than the average proportion of the deaths reported, are as follows: Measles, whooping-cough, enteric fever, diarrhœa, dysentery, malarial fever, erysipelas, puerperal septicæmia, premature birth, still-birth, scrofula and tabes, dentition, and childbirth and abortion.

7. Region of the Great Northern Lakes.

This comprises those parts of New York, Ohio, Indiana, Illinois, Michigan, and Wisconsin, which border on the Great Lakes, and it partakes to a certain extent of the characteristics of the Atlantic coast region. These large bodies of fresh water undoubtedly exert a very considerable influence upon the climate in moderating its extremes.

The mean annual temperature in the southern part of this region is from 45° to 50° Fahr., and in the northern portion from 40° to 45° Fahr. The mean annual rainfall is from 30 to 40 inches, except in northern Michigan, where it is only from 20 to 25 inches. The elevation is nowhere above 500 feet. The colored population is below 7 per cent, and the foreign population is over 30 per cent, of the whole.

The causes of death in this region, to which are attributed, more than an average proportion of the deaths reported, are as follows: Measles, scarlet fever, diphtheria, still-births, old age, cancer, convulsions, diseases of the heart, croup, peritonitis, and railroad accidents.

The causes of death in this region, to which are attributed less than the average proportion of the deaths reported, are as follows: Whooping-cough, enteric fever, dysentery, malarial fever, scrofula and tabes, consumption, dropsy, pneumonia and diseases of the respiratory system generally, dentition, diseases of the liver, urinary calculus, childbirth, and diseases of the bones and joints.

The proportion of deaths reported as due to consumption is low on the western shore of Lake Michigan in Wisconsin, and increases as we pass eastward.

8. The Interior Plateau.

This comprises that portion of the plain stretching from the base of the Appalachians eastward which includes part of Pennsylvania, Virginia, and North Carolina, and also, on the west side of the Appalachians, the plateau country of central New York and western Pennsylvania. It consists of three regions, which are not contiguous, viz., (1) the western parts of New York and Pennsylvania, (2) the southeastern corner of Pennsylvania, and (3) central portions of Virginia and North Carolina. The characteristics of the second of these regions, so far as returns of deaths are concerned, are largely due to the fact that it contains the cities of Philadelphia and Reading. These regions have little that is characteristic in climate or surface. Lying as they do between the Appalachians and the Atlantic coast region on the one hand, and the lake region on the other, they partake to a certain extent of the climate of both. The surface is broken and hilly, but nowhere rises into mountains. The group is an upland country originally covered with forests, which have been in great part cut away. It contains comparatively little water surface or swamp-land.

The mean annual temperature is from 45° to 50° Fahr. The annual rainfall is from 40 to 45 inches in that part east of the Appalachians, from 30 to 35 inches in the northern portion.

The density of population varies from 45 to 90 per square mile. The proportion of the colored population is below 7 per cent in Pennsylvania, and about 35 per cent in Virginia and North Carolina. The foreign population is about 10 per cent of the whole in Pennsylvania, and is below 1 per cent in Virginia and North Carolina.

The causes of death in this region, to which are attributed more than an average proportion of the deaths reported, are as follows: Diphtheria, debility, old age, consumption, cancer, tumor, dropsy, apoplexy, paralysis, convulsions, diseases of the heart, Bright's disease, and diseases of the kidney.

The causes of death in this region, to which are attributed less than an average proportion of the deaths reported, are as follows: Measles, whooping-cough, dysentery, malarial fever, croup, pneumonia and diseases of the respiratory system generally, and childbirth and abortion.

The proportion of deaths from consumption is high in New York, somewhat lower in Pennsylvania and Virginia, and lowest in North Carolina.

9. Southern Central Appalachian Region.

This region is a continuation of Grand Groups V and VI, passing to the southwest. It includes portions of Virginia, West Virginia, the Carolinas, Kentucky, Tennessee, Georgia, and Alabama. In Virginia and West Virginia the character of the country is very similar to that of Grand Group VI, but as we proceed southward there is a gradual rise in the ridges, and a tendency to break up into peaks, which in North Carolina develops to the highest degree, presenting in the western part of that State a complex of mountains, rising without much apparent system to heights of from 6,000 to 6,700 feet.

In Virginia and farther southward the feature which was outlined in Pennsylvania becomes very characteristic, viz., the great valley occupied in northern Virginia by the Shenandoah, farther south by the branches of the New River and the heads of the Tennessee, and in Tennessee by the river of that name. This forms a great depression which, throughout the whole region, is traversed by numberless minor ranges and ridges, while it is limited on either side by higher ranges, represented in North Carolina by the mountains of the western part of that

State, while the western boundary of the belt is the Cumberland range or plateau. In Georgia and Alabama these ranges gradually fade out and disappear. The mountains of this region rise from 1,000 to 6,700 feet above the sea, and the valleys are at elevations varying from 500 to 2,000 feet. The temperature of the habitable portions of this region varies with the altitude and the latitude, but nowhere is the mean annual temperature much higher than 55° Fahr., and it falls below 40° in the higher country. This region is covered with heavy forests of pine and hard-wood. The mean annual rainfall is from 35 to 45 inches in the northern half, and from 50 to 60 inches in the southern half. The density of population is below 45 persons to the square mile. The colored population is below 17 per cent of the whole. The foreign population is below 1 per cent, except in a few localities, and is nowhere above 5 per cent.

The causes of death in this region, to which are attributed more than an average proportion of the deaths reported, are as follows, viz.: Measles, whooping-cough, enteric fever, diarrhœa, dysentery, still-births, rheumatism, scrofula and tabes, dropsy, croup, pleurisy, urinary calculus, diseases of the bones and joints, and gunshot-wounds.

The causes of death in this region, to which are attributed less than an average proportion of the deaths reported, are as follows: Scarlet fever, diphtheria, cholera infantum, malarial fever, erysipelas, debility, old age, consumption, hydrocephalus, cancer, diseases of the nervous system, especially convulsions, diseases of the heart, pneumonia, bronchitis, Bright's disease and diseases of the kidney, and childbirth.

10. The Ohio River Belt.

This group includes those parts of Ohio, Indiana, Kentucky, and West Virginia which border on the Ohio River. It is an area of broken country, becoming more and more diversified in the upper part of the river. For the most part the rivers flow in deep, narrow valleys, bordered by high bluffs and broken hills. The area of bottom-land is limited. The mean annual temperature is from 45° to 55° Fahr. The annual rainfall is from 45 to 50 inches. The density of population is from 45 to 90 per square mile. The elevation is less than 500 feet from the mouth of the Ohio River to Cincinnati, and above this point it is from 500 to 1,000 feet. The colored population is below 7 per cent north of the Ohio River, and from 17 to 35 per cent south of that stream. The foreign population is from 5 to 20 per cent north of the Ohio River, and from 1 to 5 per cent south of it.

The causes of death in this region, to which are attributed more than an average proportion of the deaths reported, are as follows: Scarlet fever, enteric fever, cholera infantum, inanition, still-births, scrofula and tabes, consumption, diseases of the nervous system, and diseases of the bones and joints.

The proportion of deaths from consumption is comparatively high in Ohio, and somewhat lower in the rest of the group.

The causes of death in this region, to which are attributed less than an average proportion of the deaths reported, are as follows: Diphtheria, malarial fever, debility, diseases of the heart, diseases of the respiratory system, more especially croup, pneumonia, and pleurisy, diseases of the digestive system, Bright's disease, childbirth, and accidents and injuries.

11. Southern Interior Plateau.

This includes the section of the Atlantic plain which extends across South Carolina and Georgia, with the region in central Alabama and Mississippi lying between the Appalachian region and the Gulf coast belt. It is for the most part level and heavily timbered, principally with pine, a large extent of the surface being what is popularly known as "pine barrens." It has a warm climate, and during summer the temperature rises much higher than on the coast. The mean annual temperature is from 60° to 70° Fahr. The annual rainfall is heavy—from 50 to 60 inches. The density of population is from 18 to 45 persons per square mile. The elevation is for the most part below 1,000 feet. The colored population forms about 60 per cent of the whole. The foreign population is below 1 per cent, except in a few small localities.

The causes of death in this region, to which are attributed more than an average proportion of the deaths reported, are as follows: Measles, whooping-cough, enteric fever and fever unspecified, diarrhœa, dysentery and enteritis, malarial fever, puerperal septicæmia, worms, scrofula and tabes, dropsy, diseases of the respiratory system, and especially croup, pneumonia, and pleurisy, urinary calculus, dentition, childbirth, accidents and injuries, more especially burns and scalds, exposure and neglect, and gunshot-wounds.

The causes of death in this region, to which are attributed less than an average proportion of the deaths reported, are as follows: Scarlet fever, diphtheria, cholera infantum, erysipelas, debility, old age, consumption, hydrocephalus, cancer, tumor, diseases of the nervous system, diseases of the heart, bronchitis, Bright's disease, and diseases of the kidney and of the bones and joints. From consumption the proportion is very low in Georgia and Alabama, and slightly higher in South Carolina and Mississippi.

12. South Mississippi River Belt.

Along the Mississippi and Missouri Rivers lie narrow belts characterized by a considerable extent of low bottom-land with rich, deep, moist soil. All this region that borders the lower Mississippi from the neighborhood of the coast to the mouth of the Ohio is included in this group, and has very characteristic features. It includes the river counties of Kentucky, Tennessee, Missouri, Arkansas, Mississippi, and Louisiana. It is an alluvial bottom-land, lying very low with relation to the river, and subject to overflow. The drainage is poor, and there are large areas of swamp-land and stagnant water. Vegetation is very rank, being almost tropical in its luxuriance. The mean annual temperature is from 60° to 70° Fahr. The annual rainfall is from 50 to 55 inches. The density of population is from 18 to 45 persons per square mile. The elevation is between 100 and 500 feet. The colored population is about 60 per cent of the whole. The foreign population is from 1 to 5 per cent.

The causes of death in this region, to which are attributed more than an average proportion of the deaths reported, are as follows: Whooping-cough, malarial fever and fever unspecified, diarrhœa, dysentery, worms, dropsy, pneumonia, dentition, childbirth and abortion, and accidents and injuries, especially gunshot-wounds. The proportion of deaths from consumption is low in this group, with the exception of that part of Kentucky bordering on the river.

The causes of death in this region, to which are attributed less than an average proportion of the deaths reported, are as follows: Measles, scarlet fever, diphtheria, enteric fever, cholera infantum, premature birth, debility, old age, consumption, cancer, tumor, apoplexy and paralysis, diseases of the circulatory system, and especially diseases of the heart, croup, bronchitis, Bright's disease and diseases of the kidney, diseases of the bones and joints, and suicides. From consumption the proportion is very low in Arkansas, low in Louisiana, Mississippi, and Tennessee, and considerably higher near the mouth of the Ohio River in Kentucky.

13. North Mississippi River Belt.

This extends from the mouth of the Ohio to the head of the Mississippi River, including portions of Missouri, Iowa, and Minnesota on the western, and of Illinois and Wisconsin on the eastern bank. The mean annual temperature is from 40° to 45° Fahr. in the northern portion, and from 50° to 55° Fahr. in the southern portion. The annual rainfall is from 30 to 40 inches in the northern part and from 40 to 50 inches in the southern part. The density of population is from 18 to 45 persons per square mile, except in the extreme north, where it is from 6 to 18 per square mile. The elevation in the southern portion is less than 500 feet, and rises toward the north to points from 500 to 1,000 feet. The proportion of the colored population in the southern portion is from 7 to 17 per cent, decreasing toward the north, in the extreme part of which it is below 1 per cent. The foreign population varies from 10 to 35 per cent.

The causes of death in this region, to which are attributed more than an average proportion of the deaths reported, are as follows: Enteric fever, diarrhœa and dysentery, cholera infantum, malarial fever, erysipelas, septicæmia and puerperal septicæmia, inanition, still-births, debility, convulsions, tetanus and trismus nascentium, and diseases of the respiratory system, especially pneumonia.

The following are the causes of death in this region to which are attributed less than an average proportion of the number of deaths: Scarlet fever, diphtheria and whooping-cough, old age, scrofula and tabes, consumption, hydrocephalus, cancer, dropsy, diseases of the heart, Bright's disease, urinary calculus and diseases of the kidney, diseases of the bones and joints, and accidents and injuries.

14. Southwest Central Region.

This includes the northwestern part of Louisiana, the southern part of Missouri, all of Arkansas except such portions of these States as belong to the south Mississippi River belt, and central Texas. It is mainly upland, and, with the exception of parts of Texas, is heavily timbered. In Louisiana it is traversed by a narrow strip of bottom-land along the Red River. A considerable part of this region in Missouri and Arkansas is occupied by the Ozark Hills, which rise to 1,000 feet or more above sea-level, or 400 to 500 feet above the surrounding country. The mean annual temperature is from 60° to 70° Fahr. The annual rainfall is from 35 to 50 inches. The density of the population is from 6 to 18 persons per square mile, rising to from 18 to 45 in some small districts. The elevation is from 100 to 500 feet, with some peaks rising to 1,000 feet. The colored population forms from 17 to 60 per cent of the whole. The foreign population is below 5 per cent, except in southern Texas, where it rises to from 20 to 30 per cent and over.

The causes of death in this region, to which more than an average proportion of the deaths reported are attributed, are as follows: Small-pox, measles, whooping-cough, enteric fever, diarrhœa and dysentery, malarial fever, erysipelas, puerperal septicæmia, worms, diseases of the respiratory organs, more especially croup and pneumonia, diseases of the digestive organs, childbirth and abortion, and gunshot-wounds.

The causes of death in this region, to which are attributed less than an average proportion of the deaths reported, are as follows: Scarlet fever, diphtheria, cholera infantum, debility, old age, consumption, cancer, tumors, paralysis, convulsions, diseases of the heart, bronchitis, Bright's disease and diseases of the kidney, diseases of the bones and joints, and suicides. The proportion of deaths from consumption is very low throughout this region.

15. Central Region, Plains and Prairies.

This includes the plateau running across the northern part of Ohio and Indiana, and the central portions of Kentucky and Tennessee, and is essentially what is left of the eastern portion of the Mississippi Valley after taking from it other characteristic regions.

The surface is for the most part undulating, presenting neither the dead level of the prairies on the one hand, nor the broken character marking the western foot-hills of the Appalachians on the other. The timber which originally covered it has been largely cut away. The mean annual temperature is from 50° to 60° Fahr. The mean annual rainfall is from 40 to 45 inches. The density of the population is from 45 to 90 persons per square mile. The elevation is from 500 to 1,500 feet. The colored population is below 7 per cent of the whole in the northern part, and from 7 to 35 per cent in the southern. The foreign population is below 1 per cent in the southern part, and from 5 to 10 per cent in the northern portion.

The causes of death in this region, to which are attributed more than an average proportion of the deaths reported are as follows: Whooping-cough, enteric fever, scrofula and tabes, consumption, glycosuria, diseases of the nervous system, croup, pneumonia, and diseases of the bones and joints.

The proportion from consumption is high in Kentucky and Tennessee, and somewhat lower in Ohio.

The causes of death, to which are attributed in this region less than an average proportion of the number of deaths reported, are as follows: Scarlet fever, diphtheria, debility, old age, cancer, tetanus and trismus nascentium, convulsions, diseases of the circulatory system, diseases of the digestive system, Bright's disease, childbirth, and accidents and injuries.

16. The Prairie Region.

This comprises most of the State of Illinois, the southern part of Wisconsin, nearly all of Iowa, southern Minnesota, the northern part of Missouri, the eastern half of Kansas, and a considerable portion of Nebraska, with that part of Dakota lying east of the Missouri belt. Though not entirely treeless, forests cover but a small portion of the area, and these are distributed along the water-courses, on the faces of bluffs and the tops of knolls. The surface is nearly level, except where cut or scored by streams. The soil is deep, extremely fertile, and generally very retentive of moisture. Originally there were larger areas of swamp-land and standing water than at present. The mean annual temperature is from 50° to 55° Fahr. in the southern part, and 40° to 45° in the northern part. The mean annual rainfall is from 35 to 40 inches in the eastern part, and from 20 to 25 inches in the western part. The density of the population is from 18 to 45 persons per square mile in the southern and eastern sections; it is below 6 per square mile in the northern and western parts. The elevation is from 500 to 1,000 feet in the eastern portion, gradually rising to from 2,000 to 3,000 feet in the west. The colored population is below 1 per cent of the whole, except in the southern portion, where it is from 1 to 7 per cent. The foreign population in the southern part is from 1 to 5 per cent; in the northern part it is from 20 to 35 per cent.

The causes of death in this region, to which are attributed more than an average proportion of the deaths reported, are as follows: Measles, scarlet fever, diphtheria, enteric fever, cholera infantum, erysipelas, puerperal septicæmia, rheumatism, glycosuria, diseases of the respiratory system (more especially croup and pneumonia), childbirth, and diseases of the bones and joints.

The causes of death in this region, to which are attributed less than an average proportion of the deaths recorded, are as follows: Inanition, debility, consumption, hydrocephalus, cancer, dropsy, diseases of the nervous system (more especially apoplexy, paralysis, and convulsions), diseases of the heart, bronchitis, dentition, Bright's disease, and accidents and injuries.

The proportion of deaths from consumption is very low in Nebraska and Kansas, and low in Minnesota and Iowa, being somewhat higher in the remainder of the group.

17. The Missouri River Belt.

This comprises a narrow strip across Missouri, with portions of eastern Nebraska, western Iowa, and central Dakota, including in the main a broad area of bottom-land of deep, rich soil, subject to overflow in the southern portion. Higher up the river, in Dakota, we enter the sub-humid section of the country, the atmosphere being drier and the rainfall less. The mean annual temperature is from 40° to 45° Fahr. in the northern part, and from 50° to 55° in the southern part. The mean annual rainfall is from 10 to 20 inches in the northern part, and from 30 to 40 inches in the southern part. The density of population is from 18 to 45 persons per square mile in the southern portion, and it is under 2 per square mile in the northern portion.

The elevation is from 500 to 1,000 feet in the southern and central portion, and from 1,500 to 2,000 feet in Dakota. The colored population forms from 7 to 17 per cent of the whole in the southern part, and it practically sinks to nothing in the northern part. The foreign population is from 10 to 20 per cent in the southern and central portions of this region.

The causes of death in this region, to which are attributed more than an average proportion of the deaths reported, are as follows: Measles (more especially in Kansas City), diphtheria, enteric fever, diarrhœal dis-cases, malarial fever, erysipelas, puerperal septicæmia, lead-poisoning, croup (especially in Kansas City), pneumonia, and gunshot wounds.

The causes of death in this region, to which are attributed less than an average proportion of the deaths reported, are as follows: Scarlet fever (with the exception of Kansas City), whooping-cough, inanition, debility, consumption, hydrocephalus, cancer, dropsy, diseases of the nervous system, and especially apoplexy, paralysis, tetanus and trismus nascentium, convulsions, diseases of the heart, diseases of the digestive system, Bright's disease and diseases of the kidneys, and accidents and injuries.

The proportion from consumption is very low in the northern part, gradually increasing toward the south.

18. Region of the Western Plains.

This extends westward from the border of the prairie region, including parts of Texas, Kansas, Nebraska, Colorado, Wyoming, Dakota, Montana, and New Mexico. The characteristics of the prairie region are here intensified in every particular. The timber is scarce, being found only along the water-courses. The surface is a monotonous rolling expanse, covered only with sparse clumps of bunch-grass, cactus, yucca, and other plants characteristic of a dry climate. The temperature varies from 65° to 70° Fahr. in the southern part, and from 40° to 45° in the northern portion. The mean annual rainfall is from 10 to 20 inches. (The rainfall is in general below 2·5 inches; indeed, this isohyetal line may be taken in general terms as the boundary-line between this and the prairie region, although in the north the cooler climate and smaller evaporation tend to throw the boundary westward, while the reverse condition in the south tends to throw it eastward.) The extremes of temperature in this region are very great, being exceeded only in the still more arid region farther west. The density of the population is under 2 persons per square mile. The elevation is 1,500 feet in the eastern portion, rising to 4,000, 5,000, and 6,000 feet in the west. The colored population in some parts of Colorado and Kansas is from 1 to 5 per cent of the whole; in the remainder of the group it is less than 1 per cent. The foreign population is less than 1 per cent, except in some parts of Colorado, Kansas, and Nebraska, where it is from 5 to 34 per cent.

The causes of death in this region, to which are attributed more than an average proportion of the deaths reported, are as follows: Small-pox, measles, scarlet fever, diphtheria, whooping-cough, fever unspecified, enteric fever, puerperal septicæmia, pneumonia (especially in Denver), pleurisy, and accidents and injuries, more especially exposure and neglect, gunshot-wounds, and homicide.

The causes of death in this region, to which are attributed less than an average proportion of the deaths reported, are as follows: Malarial fever, still-births, debility, old age, rheumatism, scrofula and tabes, consumption, hydrocephalus, cancer, tumor, dropsy, diseases of the nervous system, more especially meningitis, apoplexy, paralysis, tetanus and trismus nascentium, and convulsions, diseases of the heart, dentition, Bright's disease and diseases of the kidney, and diseases of the bones and joints.

The proportion of deaths from consumption is very low in Texas, New Mexico, Wyoming, Kansas, and Montana; it is somewhat higher in Colorado.

19. Heavily-Timbered Region of the Northwest.

This comprises parts of Minnesota, Wisconsin, and Michigan. It is heavily timbered and well watered, containing large numbers of small lakes and considerable areas of swamp, especially in Wisconsin and Minnesota. This large water surface, together with the dense forests, tend to give to this region a moist atmosphere, although the rainfall is not great. The mean annual temperature is from 40° to 50° Fahr., and below 40° in northern Wisconsin and Minnesota. The mean annual rainfall is from 30 to 40 inches.

The density of population in Wisconsin and Michigan is from 45 to 90 persons per square mile. The elevation is from 1,000 to 1,500 feet. The colored population is below 7 per cent of the whole. The foreign population is from 20 to 30 per cent and over.

The causes of death in this region, to which are attributed more than an average proportion of the deaths reported, are as follows: Scarlet fever, diphtheria, old age, rheumatism, cancer, tumor, paralysis, diseases of the heart, childbirth, diseases of the bones and joints, and accidents and injuries.

The causes of death in this region, to which are attributed less than an average proportion of the deaths reported, are as follows: Measles, enteric fever, diarrhœa and dysentery, cholera infantum, malarial fever, puerperal septicæmia, still-births, debility, scrofula and tabes, hydrocephalus, dropsy, diseases of the nervous system, especially convulsions, diseases of the respiratory organs, more especially croup and pneumonia, diseases of the digestive system, dentition, and peritonitis.

From consumption the proportion is low in Wisconsin, and considerably higher in Minnesota and Michigan.

20. The Cordilleran Region.

This includes the region westward from the Rocky Mountains to the Cascades and Sierra Nevada, consisting mainly of the high plateau crowned by a succession of mountain-ranges forming systems of a greater or less degree of complexity. It comprises Arizona, Utah, Nevada, and portions of Colorado, Montana, Wyoming, New Mexico, California, Oregon, and Washington Territory. The climate is arid, the rainfall is small, and the extremes of temperature are great between summer and winter, and day and night. As a general thing, the mountains only are timbered, the valleys and level country being covered with herbaceous plants characteristic of an arid climate. The slopes are everywhere amply sufficient to insure good drainage, and therefore swamps and stagnant water are rare. The mean annual temperature is from 40° to 50° Fahr. in the northern and central portions, and from 60° to 65° in the southern portion. The mean annual rainfall is below 10 inches in the central and southwestern portions, and somewhat greater in the eastern and northern portions.

The density of population is below two persons to the square mile. The elevation is from 4,000 to 10,000 feet and above. The proportion of the colored population is so small as not to be worth taking into account, and the same may be said in regard to the foreign population, except in a few settlements in Colorado, Utah, Montana, and Washington Territory, where it forms from 5 to 34 per cent of the whole.

The causes of death in this region, to which are attributed more than an average proportion of the deaths reported, are as follows: Measles, diphtheria, whooping-cough, fever unspecified, erysipelas, puerperal septicæmia, alcoholism, rheumatism, dropsy, diseases of the respiratory organs, more especially pneumonia, childbirth, and accidents and injuries, more especially gun-shot-wounds, homicide, infanticide, and suicide.

The causes of death in this region, to which are attributed less than an average proportion of the number of deaths reported, are as follows: Diarrhœa and dysentery, cholera infantum, debility, old age, scrofula and tabes, consumption, hydrocephalus, cancer, tumor, diseases of the nervous system, more especially apoplexy, paralysis, tetanus and convulsions, diseases of the heart, croup, diseases of the digestive system, Bright's disease and diseases of the kidney, and diseases of the bones and joints.

The proportion of deaths from malarial fever is very low throughout the greater part of this group. The same may be said as regards consumption, heart-disease and dropsy, croup, enteric fever, and old age.

21. Pacific Coast Region.

This comprises the coast portions of Washington, Oregon, and California lying between the ranges of the Cascades and Sierra Nevada and the Pacific coast. It has a well-defined wet and dry season, the former corresponding to the winter in the eastern portion of the country, and the latter to the summer. The northern part receives much more rain than the southern part. The surface consists of a complex range of mountains known as the Coast Range, running parallel to the coast, east of which is a great valley extending from Puget's Sound to the southern part of California. This is occupied in Oregon by the Willamette and other rivers; in California by the Sacramento and the San Joaquin. East of this valley is a great uplift, represented in Washington Territory and Oregon by the Cascade Range, and in California by the Sierra Nevada. The mean annual temperature is from 55° to 65° Fahr. in the southern portion, and from 45° to 55° in the northern portion. The mean annual rainfall is above 60 inches in the north, and below 20 inches in the south. The mean density of population is below two persons per square mile, except in the vicinity of San Francisco, Los Angeles, Sacramento, and Portland. The elevation varies from the coast-line to 3,000 feet. The colored population is below 7 per cent of the whole. The foreign population in California forms 20 per cent and over of the whole; in Washington and Oregon it is from 5 to 10 per cent.

The causes of death in this region, to which are attributed more than an average proportion of the deaths reported, are as follows: Puerperal septicæmia, old age, hydrocephalus, cancer, tumor, diseases of the nervous system, especially apoplexy and paralysis, diseases of the heart, diseases of the digestive organs, especially diseases of the liver, Bright's disease, and accidents and injuries, especially gunshot-wounds, homicide, and suicide.

The proportion of deaths from consumption is high in California, and slightly lower in Oregon and Washington Territory.

The causes of death in this region, to which are attributed less than an average proportion of the deaths reported, are as follows: Measles, scarlet fever, diphtheria, whooping-cough, enteric fever, diarrhœal diseases, malarial fever, erysipelas, old age, rheumatism, scrofula and tabes, dropsy, tetanus, trismus nascentium, croup, pneumonia, pleurisy, diseases of the kidney, and diseases of the bones and joints.

In concluding this subject of the relation of special causes of death to topographical features of the country, the general result of the study may be summed up in saying that the conditions of climate, the amount of annual rainfall, the amount of low-lying and swamp land, age and sex, the distribution of the people, and the proportion of the colored and foreign population, appear to be the chief causes of the differences between the several grand groups, or between different portions of the same grand group. Except in so far as it influences climate or drainage, the geological formation of different regions does not appear to have a marked influence upon the proportion of deaths from various causes, with the exception of diseases due to impurities in the water-supply.

The following table shows the principal reported causes of death in the order of their frequency, each, except apoplexy, having caused over 1½ per cent of all the deaths from known causes:

Principal Causes of Death, in Order of their Frequency.

DEATHS FROM—	Deaths.	Per 1,000 of known causes.
Total deaths	756,893	.....
Unknown cause	37,133	.....
Consumption	91,270	126·80
Pneumonia	63,053	87·60
Diphtheria	38,143	52·99
Heart-disease	26,068	36·21
Cholera infantum	24,983	34·71
Still-born	24,876	34·56
Enteric fever	22,854	31·75
Malarial fever	20,231	28·10
Croup	17,966	24·96
Convulsions	17,844	24·79
Scarlet fever	16,388	22·76
Dropsy	14,788	20·54
Debility	14,619	20·31
Old age	14,168	19·68
Paralysis	13,907	19·32
Dysentery	13,427	18·65
Cancer	13,068	18·15
Enteritis	12,640	17·56
Diseases of the brain	12,280	17·06
Whooping-cough	11,064	15·37
Bronchitis	10,984	15·26
Inflammation of the brain	10,903	15·14
Diarrhœa	10,825	15·03
Apoplexy	9,658	13·41

The total number of deaths reported as due to consumption during the census year (1880) was 91,270, being the greatest number reported as due to any single cause of death. Of this number, 40,512 were of males, and 50,758 were of females. It is reported as causing 12,059 in every 100,000 deaths from all causes as against 14,199 in 1870, 12,453 in 1860, and 10,376 in 1850. The census figures indicate that it is more frequent in females. In the fifty large cities, out of each 1,000 deaths from known causes, it caused 131·9 in males, and 144·3 in females; and in the rural districts it caused 101·9 deaths in males and 146·6 in females. A greater mortality from this disease in the female might be expected, because women are, as a rule, *more confined to the house and more exposed to air contaminated by the products of respiration.*

The mean age at death of those reported as dying from consumption during the census year was thirty-seven years.

The following table shows the proportion of deaths reported as due to this cause at various ages:

Showing the Number of Deaths from Consumption at Each Group of Ages in each 1,000 Deaths reported as caused by this Disease.

AGES.	Males.	Females.
Under 1 year	29·86	19·28
1 year	14·76	11·23
2 years	8·73	6·96
3 years	4·69	4·11
4 years	2·98	2·87
Total under 5 years	61·00	44·46

Showing the Number of Deaths from Consumption at Each Group of Ages in each 1,000 Deaths reported as caused by this Disease—(continued).

AGES.	Males.	Females.
5–10 years	11·08	12·66
10–15 years	14·46	26·18
15–20 years	59·74	107·03
20–25 years	131·73	167·92
25–30 years	118·74	142·15
30–35 years	97·01	107·21
35–40 years	93·47	90·18
40–45 years	76·26	67·85
45–50 years	68·72	51·87
50–55 years	61·53	41·91
55–60 years	51·16	30·28
60–65 years	49·08	32·26
65–70 years	40·40	27·67
70–75 years	31·54	22·31
75–80 years	20·81	16·04
80–85 years	9·05	8·03
85–90 years	3·08	2·79
90–95 years	0·87	0·87
95 and over	0·30	0·34
Unknown	4·64	3·82

In considering this table it must be borne in mind that it does not represent the relative liability to the disease at different ages, because the decrease of the living population at the higher ages is not taken into account.

It will be seen that the great majority of the deaths from consumption occur between the ages of fifteen and sixty-five the greatest proportion in any decennium occurring between the ages of twenty and thirty.

The proportion of deaths between the ages of fifteen and thirty-five is greater in the female than in the male. If we take the group of ages from fifteen to sixty-five and compare the number of deaths reported as due to consumption with the total number of deaths from specified causes at the same group of ages, we find that the proportion is greatest in the large cities, being, per 1,000,000 deaths, for males, 307,154, and for females, 338,571, while in the rural districts it is, for males, 218,455, and for females, 298,583. At the same group of ages in those regions where distinctions of color and parentage are made, the proportions are, for whites, in each 1,000,000 deaths, males, 242,842, females, 302,046; for colored, males, 248,179, females, 326,973; for those of Irish parentage, males, 309,507, females, 375,636; and for those of German parentage, males, 249,498, females, 254,958. From these figures it would seem that the proportion of deaths from this cause in the colored race is but slightly greater than in the white, and that it is greatest of all in the Irish.

Showing for Certain Groups of Ages the Number of Deaths from Consumption, and the Proportion of Deaths from this Cause per 1,000,000 Deaths at the Corresponding Age-Groups, with Distinction of Sex, of Rural and Cities, and, for Certain Regions, of Color and Parentage.

Deaths from consumption in		Deaths.				Proportion in 1,000,000 deaths at certain ages.			
		Under 5.	5 to 15.	15 to 65.	65 and over.	Under 5.	5 to 15.	15 to 65.	65 and over.
United States..	M.	2,460	1,050	32,559	4,278	16,309	32,285	239,224	81,345
	F.	2,248	1,964	42,407	3,940	17,621	60,004	306,099	82,150
Rural..........	M.	1,887	815	22,770	3,689	17,092	30,446	218,455	82,314
	F.	1,783	1,573	32,905	3,343	18,502	57,827	298,583	85,050
Cities..........	M.	573	215	9,789	587	14,370	41,878	307,154	73,742
	F.	515	391	9,502	608	15,055	77,502	338,571	66,978
White, in 10 grand groups	M.	927	271	15,395	1,971	13,331	20,967	242,842	79,943
	F.	730	544	19,649	1,053	12,947	42,814	302,046	84,733
Colored, in 10 grand groups	M.	674	419	3,987	301	25,308	92,905	248,179	71,650
	F.	637	761	6,409	287	31,861	151,957	326,973	69,610
Irish parentage, in 14 grand groups	M.	103	63	8,897	390	15,744	40,541	309,507	98,094
	F.	89	116	4,351	339	16,139	80,780	375,636	86,707
German parentage, in 14 grand groups	M.	84	46	2,361	207	10,771	26,744	249,498	98,802
	F.	72	60	1,787	175	11,072	36,882	254,958	74,691

These figures indicate a great excess of deaths from consumption at ages under fifteen in the colored race.

The greatest proportion of the deaths reported as due to consumption appears in New England and the Middle States, the middle Atlantic coast, the Ohio Valley, the western part of Kentucky, the central part of Tennessee, and on the coast of California. The special prevalence in those counties of Mississippi bordering on the Gulf coast is, in part at least, due to the peculiar distribution of the population of this region as regards age.

The proportions indicated in Florida, northern Minnesota, California, and eastern Colorado are much too great, because of the number of deaths occurring in these localities of persons who had contracted the disease elsewhere, and who went to these places because of their supposed freedom from influences producing or aggravating the disease. The proportion of deaths is greater in the interior of Michigan and Ohio than on the lake coast, and on the Gulf coast of Texas than in the interior of that State. The regions showing the least proportion of deaths are in southern and western Georgia, central Alabama, Arkansas, Kansas, and the Western Territories; the Appalachian region also shows a low proportion as compared with the country lying on either side.

The following table indicates the relative proportion of deaths from this cause in each of the 21 grand groups, with distinction of rural and cities, and, for certain regions, of white and colored, and Irish and German parentage:

Showing for Rural and Cities, with Distinction of Sex, and for White and Colored, Irish and German Parentage, the Proportion of Deaths from Consumption in 1,000 Deaths from Known Causes.

GRAND GROUPS.	RURAL.		CITIES.		White	Colored.	Irish parentage.	German parentage.
	Male	Female.	Male	Female.				
Total	101·9	146·6	131·9	144·3	126·2	139·1	198·4	123·0
1. North Atlantic coast region......	148·7	197·2	138·0	162·8	.....	.....	231·0	140·2
2. Middle Atlantic coast region.....	130·2	168·5	136·8	148·0	140·9	175·1	212·6	147·3
3. South Atlantic coast region	76·5	101·6	138·2	145·4	88·0	105·5	....	....
4. Gulf coast region	96·0	100·9	151·2	153·2	115·8	120·6	....	....
5. Northeastern hills and plateaus..	131·0	186·1	147·1	153·2	....	....	232·0	113·4
6. Central Appalachian region	99·7	136·9	123·7	140·7	....	....	183·7	143·2
7. Region of the Great Northern lakes..........................	109·8	156·8	94·7	101·0	....	....	201·4	116·1
8. The interior plateau	116·0	166·6	142·3	160·4	138·4	170·7	171·0	105·0
9. Southern Central Appalachian region	101·5	171·0	....	....	121·3	179·3	....	....
10. The Ohio River belt.........	137·1	196·6	125·0	151·0	150·7	238·1	179·6	187·9
11. Southern interior plateau........	69·0	116·5	....	....	83·3	100·4	....	....
12. South Mississippi River belt	80·3	115·7	....	....	81·1	108·8	....	....
13. North Mississippi River belt	91·5	125·3	116·9	118·8	....	....	145·5	100·2
14. Southwest central region	59·6	84·4	....	....	70·3	77·0	....	....
15. Central region, plains and prairies	115·4	180·4	131·0	155·3	136·8	221·4	....	....
16. The prairie region..............	91·1	122·0	....	....	....	....	140·2	81·5
17. Missouri River belt..............	83·9	121·4	84·7	121·3	....	....	140·7	80·1
18. Region of the Western plains....	69·8	68·2	145·8	110·5	....	....	51·9	42·2
19. Heavily-timbered region of the Northwest	118·2	130·1	....	....	....	....	175·2	101·8
20. Cordilleran region	73·8	85·6	....	....	....	....	107·9	144·9
21. Pacific coast region.............	155·9	184·1	170·4	139·8	....	....	140·0	113·4

The States and Territories presented together for Comparison, showing the Proportion of Deaths from Consumption to Deaths from all Causes.

STATES AND TERRITORIES.	Total number of deaths from all causes.	Deaths from consumption.	Percentage from consumption.
United States	756,893	91,551	12·09
Alabama	17,929	1,729	9·00
Arizona	291	18	6·9
Arkansas	14,812	955	6·3
California	11,530	1,802	15·5
Colorado	2,547	210	8·2
Connecticut...........	8,179	1,369	16·0
Dakota	1,304	116	8·8
Delaware.............	2,212	357	16·1
District of Columbia	4,192	793	18·9
Florida	3,159	263	8·3
Georgia	21,549	1,879	8·7

The States and Territories presented together for Comparison, showing the Proportion of Deaths from Consumption to Deaths from all Causes—(continued).

STATES AND TERRITORIES.	Total number of deaths from all causes.	Deaths from consumption.	Percentage from consumption.
Idaho	323	22	6·8
Illinois	45,017	4,653	10·3
Indiana	31,213	3,943	12·3
Iowa	19,377	1,925	9·9
Kansas	15,160	1,117	7·3
Kentucky	33,718	3,733	11·5
Louisiana	14,514	1,514	10·4
Maine	9,523	1,829	19·2
Maryland	16,919	2,381	12·0
Massachusetts	33,149	5,207	15·7
Michigan	19,743	2,613	13·2
Minnesota	9,037	848	9·3
Mississippi	14,683	1,287	8·7
Missouri	36,615	3,604	9·8
Montana	336	18	5·3
Nebraska	5,930	416	7·0
Nevada	728	61	8·3
New Hampshire	5,584	866	15·5
New Jersey	18,474	2,630	14·2
New Mexico	2,436	50	2·1
New York	88,332	12,858	14·5
North Carolina	21,547	2,130	9·9
Ohio	42,610	5,912	13·8
Oregon	1,864	226	12·1
Pennsylvania	63,881	8,073	12·6
Rhode Island	4,702	691	14·8
South Carolina	15,728	1,543	9·8
Tennessee	25,919	3,767	14·5
Texas	24,735	1,622	6·5
Utah	2,314	69	3·0
Vermont	4,024	813	20·2
Virginia	24,681	3,025	12·3
Washington Territory	755	100	13·2
West Virginia	7,518	969	12·9
Wisconsin	15,011	1,681	11·2
Wyoming Territory	189	5	2·6

As has been stated, the proportions of deaths from consumption indicated in Florida, Minnesota, Colorado, and California are much too great, because of deaths in these localities of persons who had contracted it elsewhere.

The States and Territories presented together for Comparison, showing the Relation of Population per Square Mile to the Deaths from Consumption per 1,000 Inhabitants.

STATES AND TERRITORIES.	Inhabitants per square mile.	Deaths from phthisis per 1,000 inhabitants.
Maine	19·0	2·8
New Hampshire	37·0	2·4
Vermont	34·0	2·4
Massachusetts	214·0	2·9
Rhode Island	221·0	2·3
Connecticut	124·0	2·2
New York	113·0	2·5
New Jersey	144·0	2·3
Pennsylvania	94·0	1·8
Delaware	71·0	2·4
Maryland	77·0	2·4
District of Columbia	2,537·0	4·4
Virginia	35·0	1·9
West Virginia	24·0	1·5
North Carolina	26·0	1·5
South Carolina	32·0	1·5
Georgia	25·0	1·1
Florida	4·7	0·9
Ohio	77·0	1·8
Tennessee	36·0	2·4
Kentucky	40·0	2·2
Indiana	54·0	1·9
Illinois	54·0	1·4
Michigan	27·0	1·5
Wisconsin	23·0	1·2
Iowa	28·0	1·1
Missouri	31·0	1·6
Arkansas	14·0	1·1
Louisiana	19·0	1·6
Mississippi	24·0	1·0
Alabama	24·0	1·3
Texas	5·9	1·0
Kansas	12·0	1·1
Nebraska	5·8	0·9
Minnesota	9·0	1·0
Dakota	0·9	0·8
Montana	0·2	0·4
Idaho	0·3	0·6
Colorado	1·8	1·1
Wyoming	0·2	0·2
Arizona	0·3	0·4
New Mexico	0·9	0·4
California	5·4	2·0
Nevada	0·5	0·9

The States and Territories presented together for Comparison, showing the Relation of Population per Square Mile to the Deaths from Consumption per 1,000 Inhabitants—(continued).

STATES AND TERRITORIES.	Inhabitants per square mile.	Deaths from phthisis per 1,000 inhabitants.
Washington	1·0	1·3
Oregon	1·8	1·2
Utah	1·6	0·4

Showing for Fifty Cities in the United States the Number of Deaths from Consumption for each 10,000 of Population.

CITIES.	Population.	Deaths per 10,000 inhabitants.
New York, N. Y	1,206,299	35·56
Philadelphia, Pa	847,170	31·59
Brooklyn, N. Y	566,663	29·84
Chicago, Ill	503,185	16·75
Boston, Mass	362,839	33·37
St. Louis, Mo	350,518	22·93
Baltimore, Md	332,313	34·93
Cincinnati, Ohio	255,139	27·71
San Francisco, Cal	233,959	30·64
New Orleans, La	216,090	39·42
Cleveland, Ohio	160,146	17·04
Pittsburg, Pa	156,389	18·79
Buffalo, N. Y	155,134	18·16
Washington, D. C	147,293	41·95
Newark, N. J	136,508	28·42
Louisville, Ky	123,758	32·48
Jersey City, N. J	120,722	27·58
Detroit, Mich	116,340	18·48
Milwaukee, Wis	115,587	17·30
Providence, R. I	104,857	29·37
Albany, N. Y	90,758	25·89
Rochester, N. Y	89,366	23·27
Allegheny, Pa	78,682	11·18
Indianapolis, Ind	75,056	24·91
Richmond, Va	63,600	41·66
New Haven, Conn	62,882	19·24
Lowell, Mass	59,475	35·30
Worcester, Mass	58,291	25·04
Troy, N. Y	56,747	35·77
Kansas City, Mo	55,785	11·29
Cambridge, Mass	52,669	26·77
Syracuse, N. Y	51,792	26·83
Columbus, Ohio	51,647	19·55

Showing for Fifty Cities in the United States the Number of Deaths from Consumption for each 10,000 of Population—(continued).

CITIES.	Population.	Deaths per 10,000 inhabitants.
Paterson, N. J.	51,031	29·98
Toledo, Ohio	50,137	9·57
Charleston, S. C.	49,984	49·21
Fall River, Mass	48,961	27·57
Minneapolis, Minn.	46,887	17·27
Scranton, Pa	45,850	12·21
Nashville, Tenn.	43,350	31·37
Reading, Pa.	43,278	25·64
Wilmington, Del.	42,478	36·96
Hartford, Conn	42,015	28·08
Camden, N. J.	41,659	26·40
St. Paul, Minn.	41,473	11·57
Lawrence, Mass.	39,151	38·56
Dayton, Ohio	38,678	21·97
Lynn, Mass	38,274	29·00
Denver, Col.	35,629	17·40
Oakland, Cal.	34,555	17·36

Attention is here called to the relatively greater degree of immunity from consumption, as shown by the above table, in the cities of the upper lakes.

NOTE.—In this country, as elsewhere, the death-rate in the cities is larger than in the rural districts. As compared with the rural districts, the cities have been for the last twenty years gaining most in healthiness, owing to the fact that systematic sanitary work has been carried on in them to a much greater extent than in the smaller towns and villages.

The following table shows the relations of deaths reported as due to consumption in the thirty-one registration cities with relation to the month of death. It will be seen that in the aggregate the distribution of deaths from this cause is tolerably uniform throughout the year, although somewhat larger in the winter and spring months, reaching its maximum in the month of March:

Showing for Thirty-one Registration Cities the Number of Deaths from Consumption by Months for each City and the Proportion for each Month in 1,000 Deaths from Consumption for all the Thirty-one Cities.

CITIES.	Totals.	January.	February.	March.	April.	May.	June.	July.	August.	September.	October.	November.	December.
Total	19,917	1,748	1,785	1,961	1,821	1,713	1,420	1,536	1,429	1,529	1,640	1,595	1,740
Baltimore, Md	1,162	86	104	122	113	95	77	90	87	94	85	95	114
Boston, Mass	1,211	121	102	122	109	102	97	83	87	81	92	101	114
Brooklyn, N. Y	1,691	129	159	171	132	139	104	132	130	168	148	140	139
Cambridge, Mass	141	11	5	14	13	11	15	11	14	15	15	5	12
Camden, N. J	110	7	6	10	9	12	8	8	7	12	9	11	11
Charleston, S. C	246	23	25	18	17	21	23	27	13	17	18	20	24
Chicago, Ill	843	73	80	93	98	51	66	54	61	51	77	63	76
Cincinnati, Ohio	707	57	58	64	68	62	48	60	52	51	69	47	71
Cleveland, Ohio	273	25	30	27	33	21	18	21	23	24	18	18	15
Fall River, Mass	135	11	12	19	13	16	7	12	10	9	6	10	10
Indianapolis, Ind	187	11	18	12	18	13	18	23	8	16	18	20	12
Jersey City, N. J	333	34	33	32	24	22	20	25	29	34	24	34	22
Lawrence, Mass	151	15	13	20	17	10	8	11	13	10	12	10	12
Louisville, Ky	402	35	34	35	44	43	45	30	21	34	20	16	37
Lowell, Mass	210	22	22	22	22	19	16	13	14	13	22	8	17
Lynn, Mass	111	6	6	12	12	8	10	10	10	8	10	6	13
Milwaukee, Wis	200	17	20	21	15	21	20	14	16	14	14	11	17
Nashville, Tenn	136	7	15	17	13	16	10	9	7	13	7	11	11
Newark, N. J	388	48	32	33	29	38	33	34	24	29	31	31	26
New York, N. Y	4,200	405	373	386	395	353	281	330	285	329	372	308	383
New Orleans, La	852	66	70	77	78	77	49	89	56	67	94	64	65
Paterson, N. J	153	12	12	12	11	20	12	10	14	9	9	14	18
Philadelphia, Pa	2,677	249	239	280	223	249	179	210	216	191	205	208	225
Pittsburg, Pa	293	26	32	33	32	26	20	22	21	21	17	24	19
Providence, R. I	308	29	32	35	28	21	16	25	24	20	31	26	21
Richmond, Va	265	19	30	23	22	29	19	21	17	19	25	15	26
San Francisco, Cal	717	70	59	74	80	62	65	32	43	55	65	50	62
St. Louis, Mo	804	50	67	77	72	56	65	58	65	59	67	74	94
Washington, D. C	618	48	72	67	53	64	51	49	39	44	85	43	53
Wilmington, Del	157	17	9	20	16	11	14	12	11	9	13	10	15
Worcester, Mass	146	19	16	13	12	20	6	11	9	10	12	12	6
Per 1,000	......	87·76	89·62	98·45	91·42	86·00	71·29	77·12	71·74	76·76	82·34	80·08	87·36

IV.

TOPOGRAPHY AND CLIMATE OF STATES,*

AND SUMMARY FOR STATES, GROUPS, CITIES, AND FOR COUNTIES OF TEN THOUSAND POPULATION AND UPWARD, SHOWING THE NUMBER OF DEATHS FROM CONSUMPTION PER ONE THOUSAND INHABITANTS.†

State of Maine.

Topography.—Maine has an extreme length north and south of 300 miles, an extreme width of 210 miles, and an area of about 33,040 square miles or 21,145,600 acres. The surface of the State is hilly, with some considerable elevations in the center, the highest being Katahdin, 5,385 feet above the sea. North and south of the highland belt, which is an extension of the White Mountains of New Hampshire, the country is more level, and slopes gradually to the valley of the river St. John and to the ocean. The sea-coast, although only 270 miles in length in a straight line, is so deeply indented that, including the numerous islands, the shore-line is over 2,400 miles. Mount Desert is the largest of the islands, and has an area of 60,000 acres. Others of importance are Isle de Haut and

---

* The Rand-McNally Atlas, Chicago, 1887.

† Compiled from the United States Census Reports for 1880.

Deer, Fox, and Long Islands. Maine contains a great number of lakes, mostly of moderate dimensions. The largest is Moosehead, thirty-five miles long and about seven miles wide; next to this are Umbagog, Portage, Eagle, Long, Madawaska, Schoodic, Sebec, and Millinoket. The inland waters—rivers and lakes—cover a total area of 3,200 square miles, and the lakes alone of 2,300 square miles, or about one fifteenth part of the State. On the coast there are fine granite-quarries, in the interior altered Silurian and Devonian rocks, containing limestone and argillaceous slates.

Climate.—The winters are long and cold, and snow lies on the ground from three to five months. Frosts occur as early as the middle of September, and occasionally as late as June. The summers are pleasant but very short, and the temperature varies greatly during the year, the thermometer sinking sometimes as low as 25° below zero, and on a few days in July or August reaching 90° Fahr. The mean temperature for the year at Portland is about 45°, and in the extreme north at least 5° lower. At Belfast the mean for the year is 43°; highest recorded, 85°; lowest, —32°. The mean rainfall, including snow, is 43·24 inches, 60 per cent of which passes into her rivers. In the spring and early summer the sea-breezes from the Atlantic are laden with cold fogs, and the inhabitants are subject to pulmonary complaints. With this exception the general health of the State is good, and the death-rate is low.

Population, 648,936.

Inhabitants to the square mile, 19.

Deaths per 1,000 Inhabitants.

The State	2·8	Washington	2·5
Group 1	2·9	York	3·0
Androscoggin	2·8	*Group 2*	2·5
Cumberland	2·6	Aroostook	1·5
Hancock	3·4	Franklin	3·4
Kennebec	3·0	Oxford	2·7
Knox	3·0	Penobscot	2·1
Lincoln	2·7	Piscataquis	3·2
Sagadahoc	1·9	Somerset	3·7
Waldo	3·5		

State of New Hampshire.

Topography.—The length of New Hampshire from north to south is 180 miles; greatest breadth, 93 miles in the south; average breadth, about 45 miles; area, 9,305 square miles, or 5,955,200 acres. Portsmouth is the only harbor for large vessels. The White Mountains, which cover an area of 1,270 square miles, run through the northern division of the State, in a direction a little east of north, the height of the peaks ranging from 2,000 to 6,000 feet. They are broken by a number of gaps or "notches" at an average height of 1,200 feet. The general elevation of the State is about 1,200 feet above sea-level, sloping from north to south. The largest river is the Connecticut, which forms the greater part of the western boundary. Next come the Merrimac, the Androscoggin, and the Piscataqua, with their numerous tributaries. The harbor of Portsmouth is formed by an estuary known as Great Bay, and is never frozen, even in the severest winters. The principal lakes are Winnipiseogee, which has an area of seventy-two square miles; Lake Umbagog, the source of the Androscoggin

River; and the four Connecticut lakes in the north, which form the source of the river of the same name. The Isles of Shoals lie ten miles southeast of Portsmouth, and form a part of New Hampshire.

Climate.—Owing to the difference in elevation, the temperature varies considerably. In the Merrimac Valley and generally in the southern division the extremes are not so great, but the winters in the White Mountains are excessively cold, and characterized by violent winds and snow-storms.

The summers are short and hot, the thermometer sometimes rising to 98°; cold weather begins with November and lasts to the end of April, and snow lies on the ground the winter through, and on the tops of the White Mountains for eight months of the year. The precipitation of rain and snow ranges from 46 inches in the lowlands to 55 inches on the mountains. The climate is healthy.

Population, 346,991.

Inhabitants to the square mile, 37.

Deaths per 1,000 Inhabitants.

The State	2·4	*Group 2*	2·0
Group 1	2·7	Carroll	2·2
Belknap	2·0	Cheshire	1·6
Hillsborough	2·6	Coos	1·5
Merrimac	3·0	Grafton	2·1
Rockingham	2·8	Sullivan	2·5
Strafford	2·3		

State of Vermont.

Topography.—Vermont has a length north and south of about 150 miles, a breadth of from 35 to 50 miles, and an area of 9,565 square miles, or 6,121,600 acres. The Green Mountains intersect the State from north to south, and contain a number of peaks from 3,000 to 4,500 feet high. A second range, of inferior height, branches off at Killington Peak and trends northeast. There are also some detached peaks, of which Mount Ascutney, 3,320 feet high, is the most conspicuous. Lake Champlain extends for 105 miles along the western border, and receives many small rivers and creeks. The entire territory east of the mountains is drained by the Connecticut River and its numerous tributaries — the Connecticut separating Vermont from New Hampshire. The Connecticut is the only navigable river. Lake Champlain, 126 miles in length, and from forty rods to fifteen miles in width, has a depth of from fifty to nearly 300 feet, and is navigable throughout by the largest vessels. It contains a number of islands, which collectively form the county of Grand Isle, and its shores are deeply indented. The chief harbor is that of Burlington, the seat of the Vermont lumber-trade. The outlet of Lake Champlain is the Sorel or Richelieu River, which empties into the St. Lawrence.

Climate.—Vermont is subject to great extremes of temperature, although not liable to sudden changes, and the winters are severe. The annual mean temperature in the northeast is about 40°; in the south 44° to 46°; and the range of the thermometer is from 15° below to 90° Fahr., the summers being short and hot. The rainfall is greatest in the southern part and along the valley of the Connecticut, where it averages 44 inches per annum, and decreases gradually until in the northwest not more than 35 inches per annum are recorded. Much snow falls, especially on the mountains. The State is extremely healthy; miasmatic diseases are entirely unknown, pulmonary complaints much less common than in the coast States in the same latitude, and the death-rate is very low—only 1·07 per cent per annum.

Population, 332,286.

Inhabitants to the square mile, .34.

Deaths per 1,000 Inhabitants.

The State (forms one group)....	2·4	Orange	3·5
Addison	2·1	Orleans	2·1
Bennington	2·6	Rutland	2·1
Caledonia	2·6	Washington	2·6
Chittenden	2·7	Windham	1·9
Franklin	1·8	Windsor	2·4
Lamoille	2·8	Remainder of group	2·4

State of Massachusetts.

Topography.—Massachusetts has an extreme length from northeast to southwest of about 160 miles; a breadth varying from 47 miles in the western to about 100 miles in the eastern part; and an estimated area of 8,315 square miles, or 5,321,600 acres. The Elizabeth Islands, Martha's Vineyard, Nantucket, and some smaller islands lying to the south, belong to the State. The sea-coast is extremely irregular and deeply indented, and there are numerous good harbors. Of the large rivers the Merrimac alone falls into the sea within the limits of the State. The Connecticut traverses the western part of Massachusetts from north to south, and is not now navigated within the State. The Housatonic, Blackstone, and Taunton flow through Massachusetts, and the Charles and Mystic Rivers empty into Boston Bay. Nearly all the rivers afford valuable water-power, but none are navigable except the Merrimac. Two chains of the Green Mountains traverse the western division from north to south, and are known as the Taconic and Hoosac ridges; Saddle Mountain, in the extreme northwest (3,600 feet high), being the highest peak. The east and northeast divisions are hilly and broken, and the southeast is low and sandy.

Climate.—The winters of Massachusetts are severe and protracted, the summers short and warm, and the range of the thermometer from 10° below zero to 100° Fahr. The mean annual temperature is from 45° to 50°—that of spring, 43°; summer, 71°; autumn, 51°; winter, 21°. Snow falls usually during seven months, October to April. The average annual rainfall is about 42 inches.

Population, 1,783,085.

Inhabitants to the square mile, 214.

Deaths per 1,000 Inhabitants.

The State	2·9	Plymouth	2·4
Group 1	3·0	Suffolk	2·4
Barnstable	3·2	Boston (city)	3·3
Bristol	2·5	Remainder of group	7·2
Fall River (city)	2·7	*Group 2*	2·5
Essex	3·0	Berkshire	2·0
Lawrence (city)	3·8	Franklin	2·3
Lynn (city)	2·9	Hampden	2·9
Middlesex	2·4	Hampshire	2·7
Cambridge (city)	2·6	Worcester	2·8
Lowell (city)	3·5	Worcester (city)	2·5
Norfolk	2·3		

STATE OF RHODE ISLAND.

Topography.—Rhode Island has an extreme length north and south of 47 miles, an extreme width of 40 miles, and an area of 1,250 square miles, or 800,000 acres. Narragansett Bay divides it into two unequal parts, the western section being much the larger, and extending north from the Atlantic Ocean about 28 miles. The width of the bay varies from 3 to 12 miles, and it contains several islands, of which Àquidneck or Rhode Island, Canonicut, and Prudence Islands are the most important. Block Island, 10 miles to the south and at the western entrance of the bay, also belongs to this State. Rhode Island has a broken and hilly surface. Rivers are plentiful, though small, of no use for navigation, but, from their rapidity and their numerous waterfalls, of great service for manufacturing purposes. The chief rivers are the Pawtucket and Pawtuxet, emptying into Narragansett Bay, and the Pawcatuck, which falls into Long Island Sound. There are numerous small lakes in this State.

Climate.—Owing to its nearness to the sea, the climate of Rhode Island is mild and equable, resembling that of southern Massachusetts and eastern Connecticut. The mean annual temperature varies from 49° to 51° Fahr., and the annual rainfall averages about 42 inches. Owing chiefly to its mild and equable temperature, Newport has become the great fashionable summer resort of the country.

Population, 276,531.

Inhabitants to the square mile, 221.

Deaths per 1,000 Inhabitants.

The State	2·3	Newport	2·0
Group 1	2·4	Providence	2·4
Bristol	1·5	Providence (city)	2·9
Kent	1·3	Washington	2·3

State of Connecticut.

Topography.—Connecticut is the third smallest of the States, following next after Rhode Island and Delaware. Its average length is 86 miles; average breadth, 55 miles; area, 4,990 square miles, or 3,193,600 acres. The country is beautifully diversified by hills and valleys, although the scenery is less rugged than that of the States on its north. The Green Mountain range terminates in this State in a series of hills, and the highest land is about 1,000 feet above the sea-level. The Housatonic, Thames and Quinebaug, and Connecticut Valleys extend north and south through the State, and contain much of its best land. The sea-coast is over 100 miles in length, and is deeply indented by numerous bays and harbors, affording excellent anchorage for sea-going vessels. New Haven, Bridgeport, New London, Stonington, and Saybrook are the most important of these.

Climate.—Connecticut is not subject to such great extremes of temperature as Massachusetts, Vermont, and northern New York, and the climate is mild and healthy. The mean annual temperature is about 50° Fahr.; that of spring, 46°; summer, 70°; autumn, 53°; and winter, 30° Fahr. Occasionally the thermometer sinks to zero, and considerable snow sometimes falls.

The summers are warm and pleasant, and the temperature rarely exceeds 90° Fahr., even in July. The annual precipitation of rain and snow is about 48 inches.

Population, 622,700.

Inhabitants to the square mile, 124.

Deaths per 1,000 Inhabitants.

The State	2·2	*Group 2*	2·3
Group 1	2·1	Hartford	2·0
Fairfield	2·2	Hartford (city)	2·8
Middlesex	2·0	Litchfield	1·6
New Haven	2·1	Tolland	2·3
New Haven (city)	1·9	Windham	2·6
New London	2·3		

State of New York.

Topography.—The extreme length of New York east and west is 412 miles; greatest breadth from the Canadian boundary to Staten Island, 311 miles; area, 49,170 square miles, or 31,468,800 acres. The outline of the State is very irregular, and two thirds of the boundaries are formed by navigable waters, giving New York a total water frontage of 880 miles. Long Island, Manhattan, and Staten Islands are by far the most important divisions, distinct from the mainland. The narrow belt lying east of the Hudson River Valley is intersected by spurs of the Hoosac and Green Mountains, while the rolling table-lands to the west are traversed by the Blue Ridge and the Highland chains, the Catskill, Helderberg, and Adirondacks.

The chief river is the Hudson, which rises in the Adirondacks, and has a southerly course of 300 miles to New York Bay. The Alleghany and its tributaries drain the southwest, and the Susquehanna the southern central division. The Mohawk is the chief affluent of the Hudson. The State is noted for the beauty of its lakes. In the west are Chautauqua and Cattaraugus; in the central division Canandaigua, Cayuga, Onondaga, Oneida, and others, having the Oswego River for their outlet.

Climate.—No State has a greater diversity of climate than New York. The mean temperature for the State for the year is 46·50° Fahr., while in the Adirondacks the annual mean does not exceed 40°, and in the extreme south it is about 50°. The average annual precipitation is about 42 inches, the greatest fall being in the lower Hudson Valley, and the least (32 inches) in the valley of the St. Lawrence.

Population, 5,082,871.

Inhabitants to the square mile, 113.

Deaths per 1,000 Inhabitants.

The State	2·5	Essex	2·1
Group 1	3·1	Franklin	1·6
Kings	1·1	Herkimer	1·9
Brooklyn (city)	2·9	St. Lawrence	2·0
New York (city)	3·5	Warren	1·6
Queens	1·6	Remainder of group	2·6
Richmond	2·8	*Group 3*	2·0
Rockland	1·9	Delaware	1·2
Suffolk	2·0	Greene	2·7
Westchester	2·0	Orange	2·1
Group 2	2·0	Sullivan *	1·0
Clinton	2·2	Ulster	2·3

* The immediate neighborhood of Liberty, Sullivan County, is one of the best regions, particularly during the summer and autumn months, within a day's journey of New York city, for consumptives and others who

Deaths per 1,000 Inhabitants.

Group 4	1·9	Fulton	1·6
Chautauqua	1·7	Lewis	1·5
Erie	1·4	Livingston	2·3
Buffalo (city)	1·8	Madison	2·0
Genesee	1·9	Montgomery	2·0
Jefferson	1·7	Oneida	2·0
Monroe	1·9	Onondaga	1·9
Rochester (city)	2·3	Syracuse (city)	2·6
Niagara	2·0	Ontario	2·0
Orleans	2·2	Otsego	1·8
Oswego	2·0	Putnam	1·8
Wayne	1·7	Rensselaer	2·1
Group 5	2·7	Troy (city)	3·5
Albany	2·8	Saratoga	2·6
Albany (city)	2·5	Schenectady	2·8
Allegany	1·1	Schoharie	2·1
Broome	2·1	Schuyler	1·7
Cattaraugus	1·3	Seneca	2·4
Cayuga	1·9	Steuben	1·5
Chemung	1·3	Tioga	1·6
Chenango	2·0	Tompkins	2·0
Columbia	1·6	Washington	2·2
Cortland	2·1	Wyoming	2·2
Duchess	2·5	Yates	2·6

State of New Jersey.

Topography.—New Jersey has an extreme length north and south of 157 miles; a breadth of from 37 to 70 miles; and an area of 7,815 square miles, or 5,001,600 acres. The highest ground is found in the northwest, where the Blue Mountains attain an elevation of from 1,000 to 1,750 feet. The Highland Range consists of a series of hills rising 300 to 600 feet above their alternating valleys, and separated from the Blue Mount-

---

require a pure atmosphere with moderate elevation. This opinion is supported by statistics and also by clinical experience.

ains by the Kittatinny Valley, which has a width of about ten miles. The elevation of this range is from 1,000 to 1,400 feet above the sea. The Palisades of the Hudson, on the northeast, consist of rough ridges of trap formation, never exceeding 600 feet in height. The center of the State is an undulating plain, and the southern division is low and level. The Hudson forms part of the eastern border, and the Delaware River and Bay the western. The Atlantic coast-line is 120 miles long, and the water frontage on Delaware Bay is almost as great, while the Hudson River and the Raritan, New York, and Newark Bays afford splendid harbor facilities. The coast from Cape May to Sandy Hook is bordered by long strips of sand-beach inclosing considerable bodies of water connected by narrow passages. Vessels of light draught can sail round much of the coast in these protected lagoons, and thus avoid the rough sea of the Atlantic. In the northern highlands there are several small picturesque lakes, and the watering-places on the Atlantic coast, including Long Branch, Squan Beach, Atlantic City, and Cape May, are among the most popular summer resorts in the East.

Climate.—The temperature varies considerably in different parts of New Jersey, the annual mean in the northern highlands being from 48° to 50° Fahr.; while in the south, where the elevation is slight and the influence of the ocean is felt, it reaches 54°. The uplands are healthy, but marsh-fever and ague prevail in some parts of the south. The precipitation of rain and snow ranges from 41 inches at Cape May to 50 inches in the northern highlands.

Population, 1,131,116.

Inhabitants to the square mile, 144.

Deaths per 1,000 Inhabitants.

The State	2·3	Monmouth	2·3
Group 1	2·4	Ocean	2·0
Atlantic	2·0	Salem	1·5
Bergen	2·5	Union	1·8
Burlington	2·2	Remainder of group	2·3
Camden	2·0	*Group 2*	2·0
Camden (city)	2·6	Hunterdon	1·9
Cumberland	2·4	Mercer	2·7
Essex	2·1	Morris	1·7
Newark (city)	2·8	Passaic	1·5
Gloucester	1·5	Paterson (city)	2·9
Hudson	2·6	Somerset	1·4
Jersey City	2·7	Sussex	1·5
Middlesex	1·6	Warren	1·4

State of Pennsylvania.

Topography.—The greatest length of Pennsylvania east and west is 303 miles; greatest width north and south, 176 miles; mean length, 280 miles; mean breadth, 158 miles; area, 45,215 square miles, or 28,937,600 acres. That part of Pennsylvania between the Blue Mountains and the Delaware River rises from a few feet above tide-water, at Philadelphia, to nearly a thousand feet at the base of the hills, the ascent being gradual. The country is one of great beauty. The Cumberland Valley forms a part of the great depression which extends through the entire length of the Appalachian system as far south as Alabama. The mountain belt of the State is bounded east and west by the Kittatinny and Alleghany Mountains. The third great division of the State is the extensive elevated table-land which occupies about one half its area, and extends from the western slope of the Alleghanies to the Ohio border. The Susquehanna drains nearly one half the area of the State. Its chief tributary is the Juniata. The Delaware, which rises in the Catskill Mountains, in New York, is a tidal stream 132 miles from the sea, at Trenton. The Alleghany rises in the "oil country," and at Pittsburg forms a junction with the Monongahela. The Ohio, below their junction, is a great thoroughfare for steam navigation.

Climate.—The temperature of the southern and eastern divisions of the State differs considerably from that of the north, and of the portion west of the mountains. In the Alleghany, central, and northern uplands, the winters are severe and protracted, with heavy falls of snow. Along the Delaware the summer temperature often ranges from 90° to 100° Fahr., and the valleys of the Susquehanna and Juniata have a climate closely resembling that of the valley of the Rhine, the summer heat being prolonged far into October. West of the mountains the summers are shorter and hot; the winters cold. The average fall of rain and snow is from 36 to 45 inches, varying in different parts of the State. The climate is healthy; and vegetation is about a week earlier than in New York State.

Population, 4,282,891.

Inhabitants to the square mile, 94.

Deaths per 1,000 Inhabitants.

The State	1·8
Group 1	1·4
Adams	1·8
Bedford	1·4
Blair	1·2
Bradford	1·5
Cambria	1·5
Carbon	1·4
Centre	1·6
Clearfield	1·1
Clinton	1·1
Columbia	1·1
Cumberland	1·8
Dauphin	1·4
Fayette	1·4
Franklin	2·2
Fulton	2·4
Huntington	1·6
Indiana	1·1
Juniata	1·5
Lackawanna	1·6
Scranton (city)	1·2
Lebanon	1·7
Luzerne	1·1
Lycoming	1·3
Mifflin	2·7
Monroe	1·3
Mentour	0·8
Northumberland	1·3
Perry	1·3
Schuylkill	1·4
Snyder	1·6
Somerset	1·0
Susquehanna	1·6
Tioga	1·0
Union	0·9
Wayne	1·0
Westmoreland	1·2
Wyoming	1·9
Remainder of group	0·6
Group 2	2·1
Allegheny	1·5
Allegheny (city)	1·1
Pittsburg (city)	1·8
Armstrong	1·7
Beaver	1·6
Berks	1·8
Reading (city)	2·5
Bucks	2·3
Butler	1·5
Chester	2·2
Clarion	0·9
Crawford	1·5
Delaware	1·7
Elk	0·7
Erie	1·7
Greene	1·7
Jefferson	1·6
Lancaster	1·7
Lawrence	1·1
Lehigh	2·0
McKean	1·2
Mercer	1·4
Montgomery	1·9
Northampton	1·4
Philadelphia (city)	3·1
Potter	0·8
Venango	1·6
Warren	1·1
Washington	2·3
York	1·1
Remainder of group	0·9

State of Delaware.

Topography.—The State has an extreme length north and south of 96 miles; a breadth of about 36 miles on the south line and 10 miles on the north; and an area of 2,050 square miles, or 1,312,000 acres. There are no mountains in Delaware. The southern portion is almost level, and sandy, with large marshes abounding in cypress, cedar, and other trees; but the northern half is undulating, and contains some beautiful though not striking scenery. The coast is low and swampy, with salt marshes and shallow lagoons separated from the sea by sandy beaches. The water-shed is formed by a low table-land or sand-ridge running north and south near the western border, and not more than 60 to 75 feet in height. Drainage is into the Chesapeake and Delaware Bays, but the streams are unimportant, and, with the exception of Christiana Creek, only available for small craft. The tide runs up to Wilmington, where there is a depth of 18 feet of water, and Rehoboth Bay, at the mouth of Indian River, admits vessels drawing six feet of water.

Climate.—The climate is mild, and tempered by the sea-breezes. The mean annual temperature is from 51° to 53° Fahr., and the rainfall about 50 inches per annum.

In the northern division the climate is salubrious and pleasant, but in the swampy parts of the south there is considerable malaria.

Population, 146,608.

Inhabitants to the square mile, 71.

Deaths per 1,000 Inhabitants.

The State (forms one group)	2·4	Wilmington (city)	3·6
Kent	1·6	Sussex	2·0
Newcastle	1·9		

State of Maryland.

Topography.—Maryland has an extreme length east and west of 196 miles; its breadth varies from less than 10 miles in the west to about 120 miles in the eastern peninsula; while the area, not including Chesapeake Bay, which comprises 2,835 square miles, is 12,210 square miles, or 7,814,400 acres. Chesapeake Bay extends almost through the entire breadth of the State. Maryland has over 500 miles of frontage on tide-water and several navigable rivers, of which the chief are the Potomac, Patuxent, Patapsco, and Susquehanna, all of which empty into Chesapeake Bay. The extreme western part of the State is drained by the Youghiogheny, a tributary of the Monongahela. Chesapeake Bay contains numerous small islands, and its shores are indented by many bays and inlets. The peninsular section is low and sandy, and the western division, lying between Chesapeake Bay and the estuary of the Potomac, is of the same general character; but in the northwest the Blue Ridge and Alleghany Mountains attain a moderate elevation, and the country is rugged and broken.

Climate.—The climate is mild and salubrious, being modified by the vicinity of the ocean, and the State generally is healthy, although malarial diseases are not unknown in the lowlands along the bay. The mean annual temperature in the northwest is about 50° Fahr., in the central division about 56°, and at Baltimore about 54°. The rainfall averages from 45 to 50 inches per annum.

Population, 934,943.

Inhabitants to the square mile, 77.

Deaths per 1,000 Inhabitants.

The State	2·4	Montgomery	1·6
Group 1	2·6	Prince George's	2·6
Anne Arundel	2·6	Queen Anne	2 7
Baltimore	2·1	St. Mary's	1·7
Baltimore (city)	3·4	Somerset	2·5
Calvert	1·8	Talbot	2·0
Caroline	2·1	Wicomico	2·0
Carroll	1·3	Worcester	1·8
Cecil	1·9	*Group 2*	1·6
Charles	2·4	Allegany	1·7
Dorchester	2·0	Frederick	1·6
Harford	1·6	Garrett	0·9
Howard	2·4	Washington	1 7
Kent	2·8		

District of Columbia.

Topography.—Contains 70 square miles. It was originally 10 miles square, but, by the retrocession of Alexandria County to Virginia in 1846, it was reduced to its present dimensions. It borders on the Potomac River, is situated low, and is more or less swampy.

Climate.—The climate is mild but not salubrious; malarial diseases are quite common, particularly in the rural districts. The mean annual temperature is about 56° Fahr. The rainfall averages from 45 to 50 inches per annum.

Population, 177,624.

Inhabitants to the square mile, 2,537.

Deaths per 1,000 Inhabitants.

District of Columbia, total	4·4	Washington (city)	4·1
District of Columbia	5·7		

State of Virginia.

Topography.—The greatest length of Virginia east and west is about 440 miles; greatest breadth north and south, 190 miles; area, 42,450 square miles, or 27,168,000 acres. The Shenandoah, Alleghany, and Cumberland Mountains extend along the West Virginia border from Harper's Ferry to the Tennessee line. The six great topographical divisions are known as the Tidewater, Middle, Piedmont, Blue Ridge, Valley, and Appalachian sections, all of which extend across the State from northeast to southwest, and have a general trend corresponding to that of the Atlantic coast and the Appalachian range. More than three fourths of Virginia is drained by the Potomac, Rappahannock, Rapidan, York, Elizabeth, James, and their tributaries, all of which find their way at last to the Atlantic.

Climate.—Owing to the differences in elevation and situation, the climate of Virginia varies greatly in the several sections. The mean annual temperature is from 55° to 60° on the sea-coast, and from 48° to 52° Fahr. in the Blue Ridge and Appalachian districts. Observations taken at Lynchburg give the mean temperature at different seasons as follows: Spring, 51°; summer, 75°; autumn, 55°; winter, 40°; the year, 56·5° Fahr. There is an abundant rainfall, the annual precipitation being from 44 to 55 inches, most rain falling in the southeast.

Population, 1,512,565.

Inhabitants to the square mile, 35.

Deaths per 1,000 Inhabitants.

The State	1·9
Group 1	2·0
Accomac	2·1
Elizabeth City	2·6
Essex	1·5
Gloucester	1·7
Isle of Wight	0·7
King and Queen	2·8
Nansemond	0·8
Norfolk	2·9
Prince George	0·9
Southampton	1·3
Sussex	2·1
Remainder of group	1·0
Group 2	2·4
Alexandria	3·6
Amelia	2·0
Appomattox	1·7
Brunswick	2·2
Buckingham	1·8
Campbell	3·4
Caroline	1·4
Charlotte	1·9
Chesterfield	1·5
Culpeper	1·7
Cumberland	2·4
Dinwiddie	3·8
Fairfax	2·0
Fauquier	1·3
Fulvanna	2·5
Goochland	1·3
Halifax	1·8
Hanover	1·2
Henrico	1·8
Richmond (city)	4·1
Henry	1·1
Loudoun	2·0
Louisa	3·3
Lunenburg	2·7
Mecklenburg	1·5
Nottoway	2·6
Orange	3·1
Pittsylvania	1·3
Prince Edward	2·0
Spottsylvania	2·8
Remainder of group	3·1
Group 3	1·5
Albemarle	1·9
Amherst	1·5
Augusta	2·3
Bedford	1·6
Botetourt	2·4
Carroll	1·5
Floyd	1·1
Franklin	1·3
Frederick	2·3
Grayson	1·0
Lee	1·2
Madison	1·4
Montgomery	1·0
Nelson	2·0
Patrick	0·6
Roanoke	1·2
Rockbridge	1·6
Rockingham	2·1
Russell	0·9
Scott	0·8
Shenandoah	1·5
Smythe	1·2
Tazewell	0·5
Washington	1·8
Wythe	1·0
Remainder of group	1·1

STATE OF WEST VIRGINIA.

Topography.—The greatest length of the State north and south is about 240 miles; greatest breadth, 160 miles; area, 24,780 square miles, or 15,859,200 acres. West Virginia is extremely hilly. The Alleghany range on its eastern boundary contains several large peaks, and west of this range and running parallel with it, at an average distance of 30 miles, are a series of mountains scarcely inferior in height, which inclose many fertile valleys. The scenery of the mountain-regions is very fine, and forms a special attraction for tourists. A few of the smaller streams in the east are tributary to the Potomac, but the rivers generally drain into the Ohio. The western division is a rolling table-land, with a gradual slope from the mountains, where its elevation is nearly 2,500 feet, to the banks of the Ohio, 900 feet above the sea-level. The Potomac forms part of the eastern boundary. The Big Sandy, Great and Little Kanawha, Guyandotte, and Monongahela are all navigable. The slack-water navigation of the Kanawha and Monongahela is of much service to the commerce of the State, and by means of the Ohio the Southern and Southwestern cities can be reached.

Climate.—The climate is generally equable, and is not marked by any great extremes. The mean annual temperature is about 52°: that of winter, 31°; spring, 50°; summer, 72°; autumn, 54° Fahr. The average rainfall is from 43 to 45 inches, and except in the more elevated sections little inconvenience is experienced in the winters. The climate much resembles that of Virginia, and is well adapted for all agricultural purposes. The State is very healthy, the death-rate being less than 1 per cent.

Population, 618,457.

Inhabitants to the square mile, 24.

Deaths per 1,000 Inhabitants.

The State	1·5	*Group 2*	1·6
Group 1	1·4	Cabell	2·1
Barbour	1·5	Doddridge	1·5
Berkeley	2·0	Jackson	1·5
Fayette	1·7	Kanawha	1·9
Greenbrier	1·3	Marshall	1·2
Hampshire	0·8	Mason	1·6
Harrison	1·8	Ohio	1·8
Jefferson	2·2	Putnam	1·4
Lewis	1·5	Ritchie	2·5
Marion	1·6	Roane	1·2
Monongalia	1·8	Tyler	1·4
Monroe	1·3	Wayne	1·1
Preston	0·5	Wetzel	1·0
Taylor	2·0	Wood	2·1
Upshur	1·2	Remainder of group	1·4
Remainder of group	1·3		

State of North Carolina.*

Topography.—North Carolina is about 450 miles in length east and west, and has an extreme breadth of 185 miles, and an area of 52,250 square miles, or 33,440,000 acres. The west is mountainous, the center hilly, and the coast-lands low and swampy. That part of the Alleghany range which separates this State from Tennessee has a number of local names. The several ridges inclose an extensive plateau, having a general elevation of about 3,500 feet. The Black Mountains in the north-

* For special information regarding the Ashville region of this State, the reader is referred to a valuable contribution on the subject, by T. Mortimer Lloyd, M. D., which was published in the "New York Medical Journal" for April 9, 1887.

west contain Clingman's Peak, 6,940 feet, and Mount Mitchell, 6,732 feet. In the Blue Ridge are Sugar Mountain, 5,312 feet, and Grandfather Mountain, 5,900 feet. Ashville is situated in the center of western North Carolina. This region embraces an area of 6,000 square miles, having a considerable general elevation, and lies between the Blue Ridge and Smoky ranges. The geological formation is of the oldest, and the water is soft and remarkably pure. The coast-line extends over 400 miles. The coast proper is deeply indented, and contains spacious harbors at Wilmington, Beaufort, Edenton, and New Berne. Much of the land is sandy, but more of it is fertile and abounds in valuable timber. The Great Dismal Swamp extends north from Albermarle Sound into Virginia, and covers an area of about 150,000 acres. The chief river is the Cape Fear. The Roanoke and Chowan rise in Virginia, and empty into Albemarle Sound. The Tar and Neuse have their sources in the north, and flow into Pamlico Sound. The Yadkin and Catawba become, in South Carolina, the Great Peedee and the Santee.

Climate.—The climate of the State is varied. In the low country it is warm and moist; on the mountains, cool and dry. The mean annual temperature at Asheville is 55°: that of spring, 53°; summer, 72°; autumn, 54°; winter, 38° Fahr. Frosts are light and seldom occur before November, while wheat is harvested in June, and corn in the early part of September. The annual rainfall averages about 46 inches.

Population, 1,399,750.

Inhabitants to the square mile, 26.

Deaths per 1,000 Inhabitants.

The State	1·5	Guilford	1·9
Group 1	1·2	Halifax	1·7
Beaufort	1·1	Harnett	1·1
Bertie	1·3	Iredell	1·6
Bladen	1·4	Johnston	2·0
Columbus	1·5	Lincoln	1·9
Craven	2·0	Mecklenburg	2·4
Cumberland	1·5	Moore	1·0
Duplin	0·7	Nash	0·9
Greene	0·2	Northampton	1·8
Hertford	1·3	Orange	1·7
Lenoir	1·1	Person	2·5
Martin	1·5	Randolph	1·9
New Hanover	1·3	Richmond	0·7
Pasquotank	1·4	Rockingham	1·6
Pender	1·2	Rowan	2·4
Pitt	1·1	Stanley	1·1
Robeson	1·0	Stokes	1·2
Sampson	1·0	Union	0·7
Wayne	1·3	Warren	1·9
Remainder of group	1·3	Wake	2·5
Group 2	1·7	Wilson	1·1
Alamance	2·1	Yadkin	1·4
Anson	0·6	Remainder of group	2·7
Cabarrus	0·6	*Group 3*	1·0
Caswell	3·0	Ashe	0·2
Catawba	1·8	Buncombe	1·1
Chatham	2·0	Burke	1·4
Cleaveland	0·9	Caldwell	1·4
Davidson	1·4	Haywood	0·8
Davie	1·8	Henderson	0·8
Edgecombe	1·2	Madison	0·5
Forsyth	1·6	Rutherford	1·3
Franklin	2·4	Surry	1·1
Gaston	0·8	Wilkes	0·9
Granville	3·0	Remainder of group	1·0

State of South Carolina.

Topography.—South Carolina forms an irregular triangle, having the coast-line for its base, and North Carolina and Georgia for its other sides. Its extreme length east and west is about 275 miles, its greatest breadth 210 miles, and its area about 30,570 square miles, or 19,564,800 acres. The only mountains are those of the extreme northwest, the Blue Ridge. The highest peak is called Table Mountain, and has an elevation of about 4,000 feet. The coast is low; the country stretching inward for 100 miles is flat, and beyond the sand-hills which traverse what is known as the "middle country" the land rises abruptly, continuing to ascend until Table Mountain is reached. There are about 200 miles of coast-line and several good harbors, the most notable being those of Charleston and Port Royal. Along the coast are many small islands on which the "sea-island" or long-staple cotton is grown. The Savannah River forms the southwestern boundary. Other important streams are the Great Peedee, the Santee, and the Edisto; the first named being navigable for a distance of about 150 miles from the sea. There are also many small rivers, and the State is well supplied with water.

Climate.—The temperature ranges from 15° to 95° Fahr., and the mean of the different seasons is spring, 65°; summer, 80°; autumn, 68°; winter, 51°; the whole year, 67°. The average rainfall is from 46 to 50 inches, but on the Georgian border it is somewhat less. The climate is generally healthful and equable, and, aside from epidemics of yellow fever (usually confined to the seaports), the health of the State is good. Frosts seldom occur, and Aiken and some other towns have become favorite winter resorts for consumptives and other invalids, who find relief in the dry and mild climate of that region.

Population, 995,577.

Inhabitants to the square mile, 32.

Deaths per 1,000 Inhabitants.

The State	1·5	Barnwell	0·7
Group 1	1·6	Chester	2·3
Beaufort	1·1	Chesterfield	0·3
Charleston	1·4	Darlington	1·4
Charleston (city)	4·9	Edgefield	1·6
Clarendon	1 6	Fairfield	1·8
Colleton	1·2	Greenville	1·4
Georgetown	0·7	Kershaw	0·8
Hampton	0·3	Lancaster	1·3
Horry	0·5	Laurens	1·5
Marion	0·7	Lexington	1·0
Williamsburg	0·7	Marlboroug	0·7
Group 2	0·8	Newberry	2·0
Oconee	0·8	Orangeburg	1·1
Pickens	0·7	Richland	1·9
Group 3	1·5	Spartanburg	1·6
Abbeville	2·3	Sumter	1·3
Aiken	1·6	Union	1·6
Anderson	1·5	York	2·0

State of Georgia.

Topography.—The extreme length of the State north and south is 320 miles; extreme width, 254 miles; area, 59,475 square miles, or 38,064,000 acres. The surface is quite diversified. In the north are the Blue Ridge and Etowah Mountains, with other spurs of the Appalachian range. The center consists of an elevated table-land, which gradually diminishes in height until the low and swampy country near the coast and along the Florida border is reached. In the southeast corner is the Okefinokee Swamp, a series of marshes having a circuit of over 150 miles. The coast extends from Tybee Sound southwest to Cumberland Sound, a distance of about 100 miles, but owing to the irregularities and indentations the shore-line is nearly five times that length. The most important rivers falling into the Atlantic are the Savannah and Altamaha. The other principal rivers are the Ogeechee, Ocmulgee, Oconee, Satilla, Allapaha, Chattahoochee, and Flint. Many of the rivers of the mountain country are rapid, and contain picturesque cataracts. Of these, the chief are the Falls of Tallulah, in Habersham County, and Toccoa Falls, in the Tugaloo, 180 feet high; Towaliga Falls, in Monroe County; and the Amicolah Falls, which have a descent, including the rapids and the cataracts, of 400 feet in less than a quarter of a mile.

Climate.—In the north the summers are comparatively cool and the climate is healthy, but in the southern lowlands the heat is often oppressive, the thermometer sometimes reaching 110° Fahr. The winters are very mild, the temperature seldom falling below 30° Fahr. The annual mean temperature at Augusta is about 63°, and at Savannah 66°, and the rainfall is over 60 inches per annum. The swamp-lands of the southeast are unhealthy, and malarious fevers prevail at certain seasons.

NOTE.—Speaking of the presence of extensive pine-woods extending along the eastern and middle portions of this State from one end to the other, Tyndall says: "That consumptives derive benefit from air charged with the odors of pine-trees is not to be denied. . . . Continuous respiration in an antiseptic atmosphere may yet lead to results hitherto unattained."

Population, 1,542,180.

Inhabitants to the square mile, 25.

Deaths per 1,000 Inhabitants.

The State	1·1
Group 1	1·5
Chatham	3·6
Liberty	0·4
Lowndes	0·2
Screven	0·5
Remainder of group	0·7
Group 2	1·2
Bartow	1·3
Chattooga	3·2
Cherokee	0·5
Cobb	1·4
De Kalb	0·8
Floyd	1·6
Forsyth	1·1
Franklin	0·6
Fulton	2·0
Gordon	1·1
Gwinnett	1·1
Hall	0·5
Jackson	0·9
Paulding	0·5
Polk	0·9
Walker	3·1
Whitfield	2·3
Remainder of group	0·9
Group 3	0·9
Baldwin	1·2
Bibb	1·9
Brooks	1·0
Burke	0·7
Carroll	0·5
Clarke	0·9
Columbia	0·7
Coweta	1·2
Decatur	0·2
Dooly	0·7
Dougherty	1·5
Elbert	1·0
Greene	1·1
Hancock	1·1
Harris	0·8
Henry	1·0
Houston	0·4
Jasper	1·3
Jefferson	0·5
Jones	1·0
Laurens	0·7
Lee	0·6
Macon	0·7
Meriwether	1·1
Monroe	1·2
Morgan	1·0
Muscogee	1·0
Newton	1·7
Oglethorpe	1·1
Pike	0·9
Pulaski	0·7
Putnam	0·8
Randolph	0·4
Richmond	2·5
Spalding	1·1
Stewart	0 7
Sumter	0·3
Talbot	0·8
Terrell	0·3
Thomas	0·7
Troup	1·5
Upson	1·4
Walton	1·2
Warren	1·9
Washington	1·1
Wilkes	0·1
Wilkinson	0.4
Remainder of group	0·6

State of Florida.

Topography.—Florida consists of a peninsula stretching south for 350 miles, between the Atlantic and the Gulf of Mexico, and of a long, narrow strip of land running along the Gulf, to a distance of 340 miles from the Atlantic coast-line. The peninsula is about 100 miles in width, and contains nearly four fifths of the total area, which is 58,680 square miles, or 37,555,200 acres. On all sides but the north, the sea forms the boundary, and the State has 1,146 miles of coast-line, but few good harbors. The Keys and Tortugas are a chain of small coral islands south and southwest of the point of the peninsula. The most important of these is Key West, where a naval station has been established, and where there is a good harbor. The northern division of the State is generally flat and uninteresting; in the center are many patches of higher ground, which are extremely fertile; and south of latitude 28° the Everglades begin. Florida is well watered and has a number of navigable rivers, the principal ones being the St. John's, Appalachicola, Perdido, Charlotte, and Suwanee. The northern division is of limestone formation, and what is known as the "Backbone Ridge," an elevation of 150 to 175 feet, runs down the center of the peninsula, as far south as Charlotte Harbor. The southern part is of a recent coral formation, similar to that of the Keys, and the process of growth is still going on.

Climate.— The climate of this State is excellent. Frosts are rare in the north and unknown in the south, and snow never falls. The average temperature is about 72° Fahr., the thermometer rarely falling below 30° or rising above 90°, while at Key West the difference between summer and winter temperature does not exceed 15°. The atmosphere is generally dry and clear, and most of the rainfall, which is about 54 inches per annum, is in the summer months.

Population, 269,493.

Inhabitants to the square mile, 4·7.

Deaths per 1,000 Inhabitants.

The State (forms one group)	0·9	Jefferson	0·6
Alachua	0·2	Leon	0·6
Duval*	3·2	Madison	0·4
Escambia	1·8	Marion	0·4
Gadsden	0·7	Monroe	1·0
Jackson	0·2	Remainder of group	0·8

State of Ohio.

Topography.—The greatest length of Ohio east and west is 225 miles; greatest breadth, 200 miles; area, 41,060 square miles, or 26,278,400 acres. Kelley's Island and the Bass Islands in Lake Erie, north of Sandusky, belong to Ohio. The great divide which forms the water-shed passes diagonally across the State from Trumbull County in the northeast to Mercer and Darke Counties in the west, and has a general elevation of about 1,200 feet above the sea-level, rising to 1,500 feet in Logan County. The surface slopes gradually from

* The proportion of deaths from consumption indicated for this county is much too great, because of the large number of deaths occurring in Jacksonville of persons who had contracted the disease elsewhere.

the divide north and west to Lake Erie, which is 565 feet above the sea, and southwest to the Ohio River, which at Cincinnati is about 430 feet above sea-level. The Ohio is the principal river, and has a course of 430 miles on the southern and eastern border. It flows through a valley, with wooded hills rising from it to a height of 500 to 600 feet. The Muskingum, Scioto, Hockhocking, Mahoning, and Great and Little Miami are the next in importance, and all flow south into the Ohio. On the north there are smaller streams, such as the Cuyahoga, Vermillion, Huron, Chagrin, Rocky, Black, Portage, Sandusky, and Maumee, which drain into Lake Erie.

Climate.—The mean annual temperature is from 50° to 54° Fahr., the warmest section being the southwest, along the Ohio River. The climate is, as a rule, mild, but the changes of temperature are often sudden. Considerable snow sometimes falls in the north, but not in quantities to interfere with communication, or to do any damage to the crops. The mean annual precipitation of rain and melted snow varies from 36 inches on the Lake Erie shore to 47 inches in the extreme south.

Population, 3,198,062.

Inhabitants to the square mile, 77.

Deaths per 1,000 Inhabitants.

The State	1·8	Lake	1·7
Group 1	1·6	Lorain	1·8
Ashtabula	1·8	Lucas	2·3
Cuyahoga	1·7	Toledo (city)	0·9
Cleveland (city)	1·7	Ottawa	1·8
Erie	1·3	Sandusky	1·3
Geauga	2·5	Wood	1·5

Deaths per 1,000 Inhabitants.

Group 2	2·2
Adams	1·7
Athens	1·9
Belmont	1·6
Brown	3·4
Butler	2·1
Clermont	2·6
Clinton	2·7
Fairfield	1·9
Fayette	1·5
Gallia	2·6
Greene	2·6
Hamilton	3·0
Cincinnati (city)	2·7
Highland	2·3
Hocking	2·0
Jackson	1·7
Jefferson	1·5
Lawrence	1·0
Meigs	2·2
Monroe	2·1
Montgomery	2·6
Dayton (city)	2·1
Morgan	1·8
Noble	1·3
Perry	1·1
Pickaway	2·6
Pike	2·1
Preble	1·9
Ross	2·3
Scioto	2·1
Vinton	2·2
Warren	2·5
Washington	1·7
Group 3	1·6
Allen	1·5
Ashland	1·3
Auglaize	1·7
Carroll	1·0
Champaign	1·5
Clarke	2·2
Columbiana	1·2
Coshocton	1·0
Crawford	1·1
Darke	1·4
Defiance	1·9
Delaware	1·7
Franklin	1·6
Columbus (city)	1·9
Fulton	1·7
Guernsey	1·8
Hancock	1·3
Hardin	1·6
Harrison	2·0
Henry	1·1
Holmes	1·3
Huron	1·7
Knox	1·7
Licking	2·2
Logan	2·0
Madison	1·6
Mahoning	1·3
Marion	0·8
Medina	1·4
Mercer	1·5
Miami	1·9
Morrow	1·7
Muskingum	1·7
Paulding	1·8
Portage	0·9
Putnam	1·2
Richland	1·4
Seneca	1·2
Shelby	1·9
Stark	1·1
Summit	1·1
Trumbull	1·1
Tuscarawas	1·0
Union	2·1
Van Wert	1·0
Wayne	1·5
Williams	1·4
Wyandot	1·4

State of Tennessee.

Topography.—The greatest length of Tennessee east and west is 432 miles; greatest breadth, 109 miles; and area, 42,050 square miles, or 26,912,000 acres. The Appalachian Mountains separate Tennessee from North Carolina. The State is popularly divided into three sections: East Tennessee, extending from the North Carolina border to about the middle of the Cumberland table-land; Middle Tennessee, thence to the Tennessee River; and West Tennessee, occupying the territory between the Tennessee and Mississippi Rivers. The Mississippi forms the western boundary, and, with the Tennessee and Cumberland, drains about three fourths of the State. Other rivers are the Clinch, the Holston, the Forked Deer and its branches, the Big Hatchie and the Wolf River. The Tennessee and Cumberland are navigable for a considerable distance, and the other rivers afford valuable water-power.

Climate.—The climate of the State is mild and remarkably salubrious. Owing to the great elevation of the eastern division and the level plains of the west, Tennessee has a climate embracing the characteristics of every State from Canada to Mississippi. The yearly rainfall is about 46 inches, and the range of the thermometer about 45° Fahr. The mean temperature of winter is 37·87°; spring, 56·71°; summer, 74·40°; and autumn, 57·54°. As a rule the snow-fall is light, and there is but little ice. The eastern portion of Tennessee is regarded as the healthiest section of the State.

Population, 1,542,359.

Inhabitants to the square mile, 36.

Deaths per 1,000 Inhabitants.

The State	2·4	Henderson	2·8
Group 1	2·1	Henry	2·3
Anderson	2·5	McNairy	1·7
Blount	2·5	Madison	2·1
Bradley	2·4	Weakley	2·0
Campbell	1·6	Remainder of group	1·5
Carter	1·7	*Group 3*	2·3
Claiborne	0·6	Dyer	2·1
Cocke	1·7	Lauderdale	3·3
Coffee	2·4	Obion	3·7
De Kalb	2·2	Shelby	1·5
Franklin	2·4	Tipton	2·2
Grainger	2·7	Remainder of group	7·7
Greene	2·8	*Group 4*	2·8
Hamblin	2·4	Bedford	3·6
Hamilton	3·0	Cannon	2·4
Hawkins	1·9	Davidson	2·5
Jefferson	2·3	Nashville (city)	3·1
Knox	2·9	Dickson	2·8
McMinn	2·6	Giles	3·5
Marion	2·2	Hardin	1·4
Monroe	1·3	Hickman	1·1
Overton	0·9	Humphreys	2·4
Putnam	1·3	Jackson	0·7
Roane	2·4	Lawrence	1·7
Sevier	2·3	Lincoln	3·0
Sullivan	1·8	Marshall	3·6
Union	1·8	Maury	3·7
Warren	1·8	Montgomery	2·7
Washington	3·0	Robertson	4·6
White	2·5	Rutherford	2·6
Remainder of group	1·6	Smith	2·3
Group 2	2·1	Stewart	2·4
Carroll	2·3	Sumner	3·8
Crockett	2·8	Wayne	0·9
Fayette	1·7	Williamson	2·4
Gibson	2·3	Wilson	3·5
Hardeman	1·2	Remainder of group	2·5
Haywood	2·5		

State of Kentucky.

Topography.—Kentucky has an area of 40,400 square miles, or 25,856,000 acres; its greatest length east and west being 350 miles, and its greatest breadth 178 miles. The whole of Kentucky lies within the Mississippi basin, and it is essentially a table-land, sloping gradually from the southeast to the northwest. There is a mountainous area of about 4,000 square miles in the southeast, and the eastern half of the table-land has an average height of about 1,000 feet above sea-level, with ridges 500 feet higher. Kentucky is amply provided with large rivers, the Ohio and Mississippi being navigable all along its borders, and the Big Sandy, Cumberland, Licking, Kentucky, Green, Salt, Big Barren, Tennessee, and other important streams flowing through the State.

Climate.—The climate is very pleasant, though somewhat variable, and is favorable to agriculture. The average temperature is about 55° Fahr., and in winter the thermometer seldom falls below zero. The winters are of medium length, and snow does not remain long on the ground. The average annual rainfall is 50·30 inches, the greatest fall being in spring and winter; the summers and autumns being usually dry. The mean summer temperature is about 75°, and the weather, though warm, is not oppressive. The healthfulness of Kentucky is exceeded by that of but few States or countries in the same latitude.

Population, 1,648,690.

Inhabitants to the square mile, 40.

Deaths per 1,000 Inhabitants.

The State	2·2	Caldwell	2·2
Group 1	1·4	Calloway	2·8
Boyd	2·3	Casey	1·6
Carter	0·8	Christian	2·5
Clay	1·1	Clark	2·7
Floyd	1·1	Fayette	3·0
Knox	1·9	Fleming	2·4
Lawrence	1·8	Franklin	3·9
Pike	1·3	Garrard	3·4
Pulaski	2·1	Grant	2·7
Wayne	1·7	Graves	1·6
Whitley	1·4	Grayson	1·5
Remainder of group	1·3	Green	2·6
Group 2	2·3	Hardin	2·0
Boone	1·7	Harrison	2·3
Bracken	1·4	Hart	2·5
Breckenridge	1·6	Henry	3·3
Campbell	1·6	Hopkins	1·8
Crittenden	2·3	Jessamine	3·3
Daviess	1·9	Lincoln	1·6
Greenup	1·3	Logan	2·1
Henderson	1·7	Madison	2·7
Jefferson	2·6	Marion	3·7
Louisville (city)	3·2	Mercer	2·8
Kenton	2·0	Monroe	1·3
Lewis	1·4	Montgomery	2·8
McCracken	1·7	Muhlenburg	2·1
Mason	2·5	Nelson	3·0
Meade	2·6	Nicholas	3·3
Union	1·3	Ohio	1·3
Remainder of group	2·6	Owen	1·8
Group 3	1·7	Pendleton	2·5
Ballard	1·8	Scott	1·8
Hickman	1·9	Shelby	2·2
Remainder of group	1·2	Simpson	1·8
Group 4	2·4	Todd	1·3
Adair	3·5	Trigg	1·7
Allen	2·3	Warren	2·6
Barren	2·7	Washington	2·9
Bath	2·3	Webster	1·4
Bourbon	2·3	Woodford	2·2
Boyle	3·9	Remainder of group	2·3
Butler	2·0		

State of Indiana.

Topography.—The surface of Indiana is extremely level, and it has no mountains or even hills of any size. At least two thirds of the State consists of level or undulating land, and it is only along the river-valleys that the landscape is diversified and relieved by bluffs and hills. Along the Ohio, which forms the southern boundary of the State, these hills attain a height of 200 to 300 feet. The land slopes gradually from north and northeast to the southwest, and the lowest point is found at the mouth of the Wabash. The rivers mostly run southwest and empty into the Ohio. The Wabash, Kankakee, White, Maumee, and other less important streams furnish an ample supply of water-power. The State has a shore-line of 40 miles on Lake Michigan. Its extreme length, north and south, is 276 miles; average breadth, 140 miles; area, 36,350 square miles, or 23,264,000 acres. The country near the lake is sandy and low, except at Michigan City, where there are extensive hills of sand.

Climate.— The climate is somewhat variable, especially in the winter, when the winds are from the north and northwest. The mean temperature of the year is 52°; that of winter, 31°; spring, 51°; summer, 76°; and autumn, 55°; and the average rainfall is about 38 inches. Indiana is well suited for agriculture, and the fruit-trees blossom in March and the beginning of April.

Population, 1,978,301.

Inhabitants to the square mile, 54.

Deaths per 1,000 Inhabitants.

The State	1·9
Group 1	1·5
Lake	1·4
La Porte	1·5
Porter	1·6
Group 2	2·3
Clarke	2·2
Crawford	2·1
Dearborn	2·0
Dubois	2·5
Floyd	2·8
Gibson	3·2
Harrison	2·6
Jefferson	2·6
Jennings	2·1
Orange	2·0
Perry	1·4
Pike	2·0
Posey	2·0
Ripley	2·0
Spencer	2·0
Switzerland	2·1
Vanderburg	2·4
Warrick	1·6
Washington	2·3
Remainder of group	3·3
Group 3	1·0
Adams	1·3
Allen	1·3
Bartholomew	1·8
Benton	1·2
Boone	1·6
Brown	1·0
Carroll	1·5
Cass	2·1
Clay	2·0
Clinton	1·8
Daviess	1·8
Decatur	2·8
De Kalb	1·4
Delaware	2·0
Elkhart	1·0
Fayette	1·8
Fountain	2·0
Franklin	2·2
Fulton	1·6
Grant	1·3
Greene	1·7
Hamilton	1·7
Hancock	1·8
Hendricks	1·9
Henry	1·9
Howard	2·4
Huntington	1·4
Jackson	1·8
Jay	0·6
Johnson	2·4
Knox	1·4
Kosciusko	1·2
Lagrange	1·7
Lawrence	0·8
Madison	1·3
Marion	1·9
Indianapolis (city)	2·4
Marshall	1·0
Martin	2·2
Miami	1·6
Monroe	1·9
Montgomery	2·7
Morgan	2·5
Noble	2·1
Owen	2·3
Parke	1·7
Putnam	2·3
Randolph	2·0
Rush	2·8
Saint Joseph	1·6
Shelby	2·4
Steuben	1·4
Sullivan	2·4
Tippecanoe	1·8
Tipton	1·8
Vermillion	1·6
Vigo	2·7

Deaths per 1,000 Inhabitants.

Wabash	1·1	White	1·8
Warren	1·1	Whitley	1·7
Wayne	1·8	Remainder of group	2·0
Wells	1·2		

State of Illinois.

Topography.—Illinois has been very appropriately called the "Prairie State." Next after Louisiana and Delaware it is the most level State in the Union, and fully one third of its whole area is composed of high, level, grassy plains. The average elevation of these above tide-water is not over 500 feet. At Cairo, the extreme southern angle of the State, the elevation of the land is only 340 or 350 feet above the Gulf of Mexico; and at Chicago, in the northeastern section, the elevation of the business portion of the city is only 592 feet above the sea-level. The highest land in the State is in the northwestern corner, where, between Freeport and Galena, the extreme elevation is 1,150 feet above the sea. Its extreme length north and south is 385 miles; extreme width east and west, 218 miles; area, 56,650 square miles, or 36,276,000 acres. The Wabash, Ohio, and Mississippi Rivers form part of the eastern and southern and all of the western boundary-lines, thus giving the State immense frontage on navigable waters.

Climate.—The climate is generally salubrious. The upland prairies are almost entirely free from endemic diseases, and the death-rate in the cities is low. The mean annual temperature on the fortieth parallel is about 54°, that of summer 77°, and of winter 33½° Fahr.

Vegetation begins with April, and the first killing frosts occur about the end of September. The summer heat is greatly modified by the ever-present breezes, and the climate is generally favorable for out-door occupations. The mean annual rainfall is 35 inches.

Population, 3,077,871.

Inhabitants to the square mile, 54.

Deaths per 1,000 Inhabitants.

The State	1·4
Group 1	1·4
Chicago (city)	1·6
Cook	0·4
Lake	1·8
Group 2	1·6
Adams	2·3
Alexander	1·2
Carroll	1·1
Gallatin	1·5
Hancock	1·9
Henderson	1·4
Jackson	1·3
Jersey	1·4
Jo Daviess	0·7
Johnson	1·8
Madison	1·5
Massac	1·4
Mercer	1·5
Monroe	2·1
Pike	1·9
Pope	1·7
Randolph	1·6
Rock Island	1·2
Saint Clair	1·3
Union	1·2
Whiteside	1·1
Remainder of group	2·7
Group 3	1·4
Bond	1·6
Boone	1·6
Brown	1·8
Bureau	1·5
Cass	1·8
Champaign	1·2
Christian	1·0
Clark	1·9
Clay	2·0
Clinton	1·6
Coles	2·4
Crawford	1·1
Cumberland	1·2
De Kalb	1·3
De Witt	0·9
Douglas	1·0
Du Page	0·8
Edgar	2·0
Effingham	1·4
Fayette	1·5
Ford	0·8
Franklin	1·6
Fulton	1·3
Greene	1·2
Grundy	0·8
Hamilton	1·2
Henry	1·0
Iroquois	1·3
Jasper	1·8
Jefferson	1·1
Kane	1·6
Kankakee	1·2
Kendall	1·4

Deaths per 1,000 Inhabitants.

Knox	1·3	Piatt	1·5
La Salle	1·3	Richland	1·9
Lawrence	2·4	Saline	1·6
Lee	1·7	Sangamon	1·2
Livingston	1·1	Schuyler	2·2
Logan	1·8	Scott	1·0
McDonough	1·5	Shelby	1·6
McHenry	2·0	Stark	0·6
McLean	1·3	Stephenson	1·3
Macon	1·5	Tazewell	0·8
Macoupin	1·1	Vermilion	1·4
Marion	1·5	Warren	1·5
Marshall	1·4	Washington	1 0
Mason	0·9	Wayne	1·0
Menard	1·4	White	1·8
Montgomery	1·2	Will	1·2
Morgan	1·9	Williamson	1·0
Moultrie	2·0	Winnebago	1·3
Ogle	1·1	Woodford	1·0
Peoria	1·3	Remainder of group	4·8
Perry	1·2		

STATE OF MICHIGAN.

Topography.—Michigan consists of two peninsulas, known as the Upper and the Lower, and of a number of islands in Lake Michigan and Lake Superior. The total area is 58,915 square miles, or 37,705,600 acres. The two divisions of the State are dissimilar in character and configuration. The Lower Peninsula consists of plains and table-land, with occasional prairie and much timber, while the Upper is rugged and rocky, broken up by hills, which in the western portion rise to the height of 2,000 feet. The length of the Lower Peninsula from north to south is 277 miles; its greatest breadth east and west, 259 miles. Saginaw and Thunder Bays on Lake

Huron, and Grand and Little Traverse Bays on Lake Michigan, form natural harbors of great size. The surface is generally level, but there are some irregular hills in the south, and the bluffs and sand-hills bordering on Lake Michigan are from 100 to 300 feet high. The Upper Peninsula is 318 miles in length from east to west, and from 30 to 164 miles in width. The western portion of the peninsula is largely given up to mining, but in the east farming is attended with the most favorable results. The total length of the lake-shore is 1,620 miles, exclusive of the frequent bays and inlets, and the State contains numerous rivers and small lakes. The principal islands are Isle Royale and Grand Island, in Lake Superior; Marquette, Mackinaw, and Bois Blanc, in Lake Huron; and the Beaver, Fox, and Manitou groups in the northern part of Lake Michigan.

Climate.—Michigan is a State of great climatic differences. The climate of the southern portion is comparatively mild, but that of the northern is cold and rigorous in winter. The mean annual temperature at Detroit for nineteen years was 47·25° Fahr., and at the Sault Ste. Marie 40·37°—a difference of 7°. The peach-orchards and vineyards along the entire fruit-belt, from St. Joseph to Grand Traverse Bay, prove that the climate is not so severe as to interfere with fruit-raising. The average annual rainfall at Detroit is 30·07 inches, and at Sault Ste. Marie 31·35 inches. The mean summer temperature at the two points named was 67·60° and 62° respectively. At Marquette, in the Upper Peninsula, the annual mean was 38·3°, and the average rainfall 23·46 inches. The climate is healthy, and the death-rate low.

Population, 1,636,937.

Inhabitants to the square mile, 27.

Deaths per 1,000 Inhabitants.

The State	1·5	Branch	1·4
Group 1	1·4	Calhoun	1·4
Allegan	1·4	Cass	1·6
Bay	1·3	Clinton	1·3
Berrien	1·8	Eaton	1·9
Houghton	0·5	Genesee	1·6
Huron	0·4	Gratiot	1·0
Macomb	1·9	Hillsdale	1·8
Manistee	1·1	Ingham	1·3
Marquette	0·7	Ionia	1·3
Mason	1·8	Isabella	2·2
Menominee	0·5	Jackson	1·3
Monroe	1·4	Kalamazoo	1·8
Muskegon	1·3	Kent	1·9
Oceana	2·0	Lapeer	1·3
Ottawa	1·5	Lenawee	1·6
Saginaw	1·4	Livingston	1·8
Saint Clair	1·4	Mecosta	1·2
Sanilac	1·3	Montcalm	0·7
Tuscola	1·1	Newaygo	1·3
Van Buren	1·8	Oakland	1·6
Wayne	1·6	Osceola	1·3
Detroit (city)	1·8	Saint Joseph	1·2
Remainder of group	1·4	Shiawassee	1·4
Group 2	1·6	Washtenaw	1·6
Barry	1·7	Remainder of group	4·9

State of Wisconsin.

Topography.—The scenery of Wisconsin is more diversified than that of the States contiguous to it, although its general character is that of a large plain. The plain is from 600 to 1,500 feet above the level of the sea, the highest lands being those at the sources of the rivers tributary to Lake Superior, which, near the Montreal River, are 1,700 feet above the ocean. The Mississippi, Fox, and Wisconsin Rivers have a considerable descent while passing through or along the boundary of the State, thus furnishing valuable water-power for mechanical purposes. In the southwest part of the State there are a number of elevations known as "mounds," ranging from 1,200 to 1,700 feet above the sea-level, and the cliffs on the east shores of Green Bay and Lake Winnebago form a bold and commanding ridge, from which there is a gradual slope to Lake Michigan, 589 feet above the sea. Besides the Great Lakes—Superior on the north and Michigan on the east—there are numerous bodies of water in the central and northern parts of the State. These lakes are from 5 to 30 miles in extent, with high, picturesque banks, and, as a rule, deep water. From these, many rivers take their rise, a number having beautiful cascades or rapids, and flowing through narrow, rocky gorges, or "dells," the scenery of which has become famous. Wisconsin has an endless variety of beautiful scenery. The four lakes which surround Madison; the Dells, near Kilbourn City; the weird beauty of Devil's Lake, which in the mystery of its origin rivals Lake Tahoe; and the calm peace which reigns at Geneva Lake, all possess attractions for summer tourists. The greatest length of Wisconsin north and south is 300 miles; greatest breadth east and west, 260 miles; area, 56,040 square miles, or 35,865,600 acres.

Climate.—Although Wisconsin is far north, the cold

of winter is tempered by the vicinity of Lake Michigan, and the excessive heat of the short summers is modified by the breezes from that body of water and from Lake Superior. The mean annual temperature of the southern section is 46° Fahr.; that of winter, 20°; of spring and autumn, 47°; and of summer, 72°. The winters are uniform, with heavy snows in the north preceding the hard frosts, and in the south snow often falls to a depth of 18 inches. Spring is backward, summer short and hot, and the autumn mild and pleasant. The northern part of Lake Michigan is frozen over most winters, but the ice never extends so far south as Milwaukee. The Milwaukee River is frozen up from the end of November to about the middle of March, or an average of 100 days. The prevailing winds in autumn and winter are from the west, in summer from the southwest, and in spring from the northeast, and the climate is healthful and invigorating. The annual quantity of rain and melted snow averages about 32 inches.

Population, 1,315,497.

Inhabitants to the square mile, 23.

Deaths per 1,000 Inhabitants.

The State	1·2	Racine	1·2
Group 1	1·3	Sheboygan	1·2
Brown	1·4	*Group 2*	1·0
Door	0·2	Buffalo	0·5
Kenosha	2·4	Crawford	1·2
Kewaunee	1·2	Grant	1·0
Manitowoc	1·0	La Crosse	1·0
Milwaukee	0·4	Pierce	1·0
Milwaukee (city)	1·7	Saint Croix	1·0
Ozaukee	1·3	Trempealeau	0·9

Deaths per 1,000 Inhabitants.

Vernon	0·9
Remainder of group	1·9
Group 3	1·3
Calumet	1·1
Columbia	1·6
Dane	1·3
Dodge	1·2
Fond du Lac	1·5
Green	1·5
Green Lake	1·5
Iowa	1·3
Jefferson	1·3
Juneau	1·6
Lafayette	1·0
Monroe	1·4
Richland	1·4
Rock	1·3
Sauk	0·9
Walworth	1·4
Washington	1·2
Waukesha	1·1
Waushara	1·2
Winnebago	1·5
Remainder of group	1·6
Group 4	1·0
Chippewa	1·0
Clark	0·6
Dunn	1·7
Eau Claire	1·0
Jackson	0·9
Marathon	0·4
Outagamie	1·1
Polk	1·0
Portage	1·2
Shawano	1·2
Waupaca	1·8
Remainder of group	0·7

State of Iowa.

Topography.—Nearly the whole State consists of gently undulating prairie, and is destitute of mountains or even hills of any size. There are some bluffs on the river-margins, and in the northeastern part the surface is more elevated and the scenery more diversified. The country is well watered and extremely beautiful, abounding with natural meadows and verdant plains. The streams, without exception, flow into one or the other of the great boundary rivers, and give unrivaled natural drainage for the whole State. In the northern portion there are numerous small, beautiful lakes, which are a part of the system extending northward into Minnesota. Its general extent north and south is 208 miles, and east and west about 300 miles; and its area is 56,025 square miles, or 35,856,000 acres, being almost exactly the same as that of Illinois. The highest point in the State is at Spirit Lake, in the northwest part, which is 1,650 feet above the sea-level, and there is a gradual slope thence to the southeast, until, at the mouth of the Des Moines River, the elevation is only 444 feet.

Climate.—It is a healthy region—malarial, epidemic, and endemic diseases being rare. The winters are severe, owing to the prevalence of north and northwest winds, which sweep at will over the prairies, but they are not unhealthy. In summer the constant breezes relieve the heat of the season. The mean annual temperature is about 48° Fahr.; that of summer, 70½°, and of winter, 23½°; and the temperature is seldom lower than 10° or higher than 90°. The mean annual rainfall for thirty years was 44·27 inches; the least 23·35 inches, and the greatest 74·49 inches. Taking the whole year, the climate is moderate, and favorable for agriculture; fruit-trees blossom early in May, and wheat ripens in August.

Population, 1,624,615.

Inhabitants to the square mile, 28.

Deaths per 1,000 Inhabitants.

The State	1·1	Muscatine	1·4
Group 1	1·0	Scott	0·8
Allamakee	1·3	*Group 2*	1·1
Clayton	1·1	Adair	0·7
Clinton	1·0	Adams	0·7
Des Moines	1·6	Appanoose	1·3
Dubuque	1·2	Benton	1·7
Jackson	1·1	Black Hawk	1·0
Lee	2·0	Boone	1·1
Louisa	1·9	Bremer	0·9

Deaths per 1,000 Inhabitants.

Buchanan	1·2	Madison	1·6
Butler	0·9	Mahaska	1·3
Carroll	0·6	Marion	1·1
Cass	0·6	Marshall	1·3
Cedar	1·0	Mitchell	1·7
Cerro Gordo	0·9	Monroe	1·7
Chickasaw	0·8	Montgomery	0·5
Clarke	1·3	Page	0·9
Crawford	0·6	Polk	1·4
Dallas	0·9	Poweshiek	0·7
Davis	0·9	Ringgold	0·5
Decatur	1·3	Shelby	1·4
Delaware	1·4	Story	0·8
Fayette	1·5	Tama	0·9
Floyd	1·1	Taylor	1·0
Franklin	1·1	Union	1·0
Greene	0·9	Van Buren	1·5
Grundy	0·5	Wapello	1·7
Guthrie	1·2	Warren	1·1
Hamilton	0·5	Washington	1·3
Hardin	1·1	Wayne	0·7
Henry	2·1	Webster	2·0
Howard	0·4	Winneshiek	1·0
Iowa	1·6	Remainder of group	0·4
Jasper	1·4	*Group 3*	0·8
Jefferson	2·0	Fremont	0·7
Johnson	1·2	Harrison	1·0
Jones	2·0	Mills	0·4
Keokuk	1·2	Pottawattamie	0·9
Linn	1·2	Woodbury	1·6
Lucas	0·7	Remainder of group	0·7

State of Missouri.

Topography.—Missouri has a length north and south of 275 miles; an average breadth of about 245 miles; and an area of 69,415 square miles, or 44,425,600 acres. That part of the State which lies north of the Missouri River consists of rolling or level prairies, with deep river-valleys, and a general slope from northwest to southeast. The southern division, which is much the larger of the two, is more broken and rugged, with a number of hills ranging from 500 to 1,000 feet in height, and mountain-ranges (the Iron Mountains and Ozark Mountains) in the extreme south. The uplands cover more than half of this section; and west of the Ozark region the prairies are undulating, and the valleys of the rivers both wide and deep. The principal rivers are the Mississippi (which washes the entire eastern boundary nearly 500 miles) and the Missouri. The Missouri has numerous tributaries within the State, chief of which are the Osage and Gasconade.

Climate.—The range of temperature is great, and the climate is subject to frequent changes. The summers are hot and the winters severe, even the largest rivers being sometimes frozen entirely over. The mean annual temperature of the central part is 55°; that of spring, 56°; summer, 76°; autumn, 55°; and winter, 39° Fahr. Southerly winds predominate, and the annual rainfall is about 32 inches, the greatest precipitation being in May.

Population, 2,168,380.

Inhabitants to the square mile, 31.

Deaths per 1,000 Inhabitants.

The State	1·6
Group 1	2·0
Bollinger	1·1
Cape Girardeau	2·2
Clarke	1·4
Jefferson	0·9
Lewis	1·5
Lincoln	2·1
Marion	1·7
Perry	2·3
Pike	2·3
Saint Charles	1·9
Sainte Genevieve	0·8
Saint Louis	1·4

Deaths per 1,000 Inhabitants.

Saint Louis (city)	2·2
Stoddard	1·7
Ralls	1·3
Remainder of group	1·0
Group 2	1·1
Barry	1·2
Barton	0·3
Bates	1·6
Benton	1·0
Cass	1·1
Cedar	1·3
Crawford	1·8
Dade	0·7
Dent	1·2
Greene	1·6
Henry	1·0
Jasper	1·4
Johnson	0·9
Laclede	1·1
Lawrence	0·9
Morgan	1·8
Newton	1·7
Pettis	1·2
Phelps	1·3
Polk	1·3
Saint Clair	0·4
Texas	0·9
Vernon	0·9
Washington	1·0
Webster	1·0
Remainder of group	1·0
Group 3	1·6
Adair	1·3
Audrain	1·5
Caldwell	1·2
Clinton	1·7
Daviess	2·0
De Kalb	1·1
Gentry	1·2
Grundy	1·5
Harrison	1·7
Knox	1·8
Linn	1·7
Livingston	1·3
Macon	1·6
Mercer	1·7
Monroe	2·2
Nodaway	1·3
Putnam	1·0
Randolph	1·9
Schuyler	1·3
Scotland	1·4
Shelby	1·7
Sullivan	1·0
Remainder of group	2·3
Group 4	1·7
Andrew	1·8
Atchison	1·2
Boone	2·4
Buchanan	1·4
Calloway	1·6
Carroll	0·8
Chariton	0·7
Clay	1·4
Cole	1·9
Cooper	2·2
Franklin	1·0
Gasconade	1·2
Holt	1·9
Howard	2·9
Jackson	1·3
Kansas City	1·1
Lafayette	2·1
Moniteau	2·3
Montgomery	2·2
Osage	2·1
Platte	2·0
Ray	2·1
Saint François	0·9
Saline	1·7
Warren	1·0

State of Arkansas.

Topography. — Arkansas has an extent north and south of 280 miles; a breadth from east to west of from 170 to 250 miles; and an area of 53,850 square miles, or 34,464,000 acres. The eastern portion of Arkansas is low and flat, but toward the west the land gradually rises and becomes somewhat hilly. The Ozark Mountains in the northwest are little more than hills, seldom attaining an elevation of over 2,000 feet, and the extreme west consists of an elevated plain, with a gradual ascent toward the Indian Territory. The most important river is the Arkansas, which rises in the Rocky Mountains, flows through Colorado and Kansas, and thence southeast through the Indian Territory and Arkansas, to its junction with the Mississippi at Napoleon. It has a course within the State of 500 miles. The Red, St. Francis, White, and Ouchita Rivers are all large streams, and of much service in commerce. The Mississippi, here of great width, washes the eastern boundary of Arkansas, and gives it an additional water frontage of nearly 400 miles. All parts of the State are finely timbered. There are extensive pine-forests; also an abundance of oak, hickory, walnut, linn, locust, cypress, cedar, and many other useful trees. The Hot Springs form one of the most remarkable natural phenomena to be found in this country. They are of great medical value, and around these famous springs a town has grown up.

Climate.— The temperature is moderate, ranging

from 15° to 100° Fahr., and frosts are seldom known. The mean annual temperature is about 63°; that of winter, 45·82°; summer, 80°; and the thermometer rises above 90° only during July and August. The rainfall varies from 40 to 55 inches per annum, the heaviest fall being in the southeastern part of the State, and the least in the northwest. In general the climate is very pleasant and healthful. The northwestern portion of the State bears a high reputation as a sanitary resort.

Population, 802,525.

Inhabitants to the square mile, 14.

Deaths per 1,000 Inhabitants.

The State	1·1	Hempstead	0·6
Group 1	1·4	Independence	1·4
Chicot	0·2	Izard	1·9
Jefferson	1·3	Jackson	1·0
Lee	1·3	Johnson	0·6
Phillips	1·8	Logan	0·9
Remainder of group	1·4	Lonoke	1·3
Group 2	1·1	Madison	0·7
Ashley	1·2	Nevada	0·8
Benton	1·0	Ouachita	0·8
Boone	0·9	Pope	2·0
Carroll	0·9	Pulaski	2·6
Clark	0·5	Randolph	1·2
Columbia	0·8	Sebastian	1·0
Conway	0·7	Union	0·5
Crawford	1·0	Washington	0·9
Drew	1·2	White	1·9
Faulkner	1·0	Yell	1·2
Franklin	1·2	Remainder of group	1·0

State of Louisiana.

Topography.—Louisiana has an extreme length east and west of 300 miles; the greatest breadth is 240 miles; area, 48,720 square miles, or 31,180,800 acres. It is low-lying, and much of the southern part is only a few feet above the sea-level. Hills there are none, except in the northwest, where there are some low ranges, never exceeding 200 feet in height; and on the east bank of the Mississippi, where the bluffs rise gradually between Baton Rouge and Natchez to the height of 200 feet. The coast-line extends over 1,200 miles, and is exceedingly irregular. Few States, if any, are so well watered, and many of the streams are navigable. The Mississippi flows for 800 miles through or on the borders of Louisiana, and reaches the sea by means of numerous branches, forming an extensive delta. The Red, Atchafalaya, Amite, Pearl, and Washita Rivers are all navigable for considerable distances. In many cases the rivers expand into large bayous or lakes. Of these, the principal are Lakes Pontchartrain, Borgne, Verret, Grand, Sabine, White, Black, Bistineau, Catahoula, Maurepas, and Washa. There are many bays and inlets on the coast, and numerous small islands in the Gulf of Mexico.

Climate.—The mean annual temperature at New Orleans is 68°, and at Shreveport, in the northwest, 64° Fahr., and the rainfall ranges from 50 to 65 inches, most of it being in spring and summer. The summers are protracted and occasionally very hot, and the winters are colder than those of the Atlantic States in the same latitude, owing to the free sweep which the northern winds have over the State. The climate is favorable to the growth of all agricultural productions, but can not be considered healthy, at least for persons who have not become acclimated. In 1853, 1867, and again in 1878, yellow fever prevailed as an epidemic in New Orleans and other cities, causing great loss of life, and an almost entire suspension of business.

Population, 939,946.

Inhabitants to the square mile, 19.

Deaths per 1,000 Inhabitants.

The State	1·6	East Carroll	0·9
Group 1	2·1	East Feliciana	2·2
Ascension	0·7	Madison	0·7
Assumption	0·8	Point Coupée	0·8
Calcasieu	0·4	Tensas	1·3
East Baton Rouge	1·9	West Feliciana	0·9
Iberia	0·9	Remainder of group	0·9
Iberville	0·8	*Group 3*	0·8
Jefferson	1·6	Bienville	0·2
Lafayette	0·4	Bossier	0·9
Lafourche	0·6	Caddo	1·1
New Orleans (city)	3·9	Catahoula	0·6
Plaquemines	1·5	Claiborne	0·6
Saint James	0·4	De Soto	0·7
Saint Landry	0·7	Lincoln	0·7
Saint Martin	0·8	Morehouse	0·8
Saint Mary	1·4	Nachitoches	0·7
Terrebonne	0·9	Ouachita	2·3
Remainder of group	0·8	Rapides	0·6
Group 2	1·0	Union	0·8
Avoyelles	0·2	Webster	0·3
Concordia	0·8	Remainder of group	0·5

State of Mississippi.

Topography.—The extreme length of Mississippi, north and south, is 332 miles; extreme breadth, 189 miles; average breadth, 142 miles; area, 46,810 square miles, or 29,958,400 acres. The surface is undulating, with an elevation in the north and northeast of from

400 to 700 feet, some of the hills rising 200 to 300 feet above the adjoining country, and has a general slope south and southwest. In the north, from Vicksburg to the Tennessee border, is the Mississippi bottom, a low, flat, swampy country, though extremely fertile. The central and southern divisions are generally hilly, with an average elevation of from 100 to 200 feet above sea-level. There are extensive marshes in the extreme south. The actual coast-line on the Gulf of Mexico is about 90 miles, but owing to irregularities the measurement is almost doubled. There are harbors at Biloxi, Mississippi City, and on the Bay of St. Louis, but the depth of water is not great. Cat and Ship Islands, and some half a dozen other small, sandy islands, lie about ten miles off the coast. The drainage of the State is by the Mississippi and its tributaries, the Big Black, Yazoo, and Bayou Pierre, and by the Pearl and Pascagoula Rivers, directly into the Gulf. The Tennessee forms a part of the boundary in the northeast, and the Tombigbee rises in the same section and flows into Alabama.

Climate.—The climate is very mild, and snow and ice are unknown. The summers are long and hot, July and August being the warmest months, and having a mean temperature of 82° to 85° Fahr. The mean annual temperature is from 65° to 66° Fahr.; and the rainfall varies from 45 to 48 inches in the north, and from 55 to 60 inches yearly on the Gulf coast. The higher lands are healthy enough, but along the rivers malarial diseases are frequent, and occasionally assume an epidemic character, resulting in great loss of life.

Population, 1,131,597.

Inhabitants to the square mile, 24.

Deaths per 1,000 Inhabitants.

The State	1·0	Noxubee	1·5
Group 1	1·1	Oktibbeha	0·7
Group 2	1·1	Panola	1·2
Alcorn	1·9	Pike	0·7
Amite	0·7	Pontotoc	1·4
Attala	0·7	Prentiss	1·3
Benton	1·3	Rankin	0·8
Calhoun	0·8	Scott	0·3
Carroll	1·4	Tate	1·0
Chickasaw	1·6	Tippah	1·3
Clarke	0·3	Union	1·9
Clay	1·5	Winston	1·0
Copiah	0·9	Yalobusha	1·9
Grenada	1·6	Remainder of group	0·7
Hinds	1·3	*Group 3*	1·1
Holmes	0·6	Adams	1·2
Itawamba	0·9	Bolivar	1·1
Jasper	0·4	Claiborne	0·4
Kemper	0·8	Coahoma	1·1
Lafayette	1·5	De Soto	1·7
Lauderdale	0·5	Issaquena	0·4
Leake	0·2	Jefferson	0·6
Lee	1·3	Le Flore	1·8
Lincoln	0·3	Tallahatchie	1·3
Lowndes	1·5	Warren	1·3
Madison	1·6	Washington	0·7
Marshall	1·9	Wilkinson	1·4
Monroe	1·1	Yazoo	1·1
Montgomery	0·8	Remainder of group	0·6
Newton	0·8		

State of Alabama.

Topography.—Alabama is 330 miles in length, and, on the average, 154 miles in breadth; and has an area of 52,250 square miles, or 33,440,000 acres. In the north-east the country is rugged and uneven, and the southern extremity of the Alleghany Mountains extends thence west, forming the dividing line between the head-waters of the Tennessee and the rivers which flow south to the Gulf of Mexico. The slope from this to the south is gradual, with rolling prairies in the center of the State, and the extreme southern portion is flat, and but slightly elevated above the sea-level. There are about 60 miles of sea-coast, including Mobile Bay, the finest harbor on the Gulf. The Mobile River is formed by the junction of the Alabama and Tombigbee; and the Chattahoochee, Coosa, and Tennessee all have a part of their course in Alabama.

Climate.—Although Alabama lies within seven degrees of the tropics, its climate is not unpleasant, the mean annual temperature being about 63° Fahr. In the northern and more elevated sections the temperature is moderated by the sea-breezes, and seldom exceeds 95°, except in July, when the thermometer has been known to record 104°. In the winter months the range is from 20° to 80°, and in spring from 25° to 90° Fahr. Snow very seldom falls, and ice is almost unknown. The rainfall varies from 46 to 49 inches per annum.

Population, 1,262,505.

Inhabitants to the square mile, 24.

Deaths per 1,000 Inhabitants.

The State	1·3	Calhoun	2·1
Group 1	2·7	Cherokee	1·2
Mobile	3·1	Cleburne	0·5
Remainder of group	1·2	Colbert	1·8
Group 2	2·1	De Kalb	1·7
Blount	1·6	Etowah	1·9

Deaths per 1,000 Inhabitants.

Jackson	2·2	Dallas	1·4
Jefferson	1·1	Elmore	1·0
Lauderdale	2·6	Fayette	0·4
Lawrence	1·9	Green	0·7
Limestone	3·1	Hale	1·4
Madison	3·3	Henry	0·1
Marshall	2·4	Lamar	0·5
Morgan	2·8	Lee	1·1
Saint Clair	1·2	Lowndes	1·5
Shelby	1·1	Macon	0·5
Remainder of group	2·5	Marengo	1·1
Group 3	0·9	Monroe	0·3
Autauga	1·6	Montgomery	1·1
Barbour	0·6	Perry	0·7
Bullock	0·7	Pickens	1·0
Butler	0·4	Pike	0·6
Chambers	1·0	Randolph	0·7
Chilton	0·6	Russell	0·8
Choctaw	1·0	Sumter	1·5
Clarke	0·7	Talladega	1·1
Clay	0·8	Tallapoosa	0·6
Conecuh	0·5	Tuscaloosa	0·6
Coosa	0·4	Wilcox	1·0
Crenshaw	0·4	Remainder of group	1·1
Dale	0·3		

State of Texas.*

Topography.—Greatest length of the State, 825 miles; greatest breadth, 740 miles; area, 265,780 square miles, or 170,099,200 acres. Its sea-coast, of about 400

* Loomis ("Practical Medicine") says: "The extraordinarily dry belt of country which runs northward from San Antonio, Texas, has begun to endanger the supremacy of Florida as a winter health resort for the consumptive. That this belt offers some climatic advantages for weak lungs over the mild but rather humid air of Florida can not be doubted."

Boerne, Kendall County—the principal health resort for consumptives in this State—is about 30 miles northwest of San Antonio; it has an elevation of 1,300 feet, an average mean temperature for the winter of 56°, miles, is irregular and bordered by many small islands. The mountains of the district lying between the Pecos and the Rio Grande attain an elevation of from 4,000 to 6,000 feet; the west and northwest sections are an elevated table-land, and thence the slope is gradual to the sea, the south and southeast divisions being flat and low. The largest and most accessible bay is that of Galveston, which extends inland 35 miles from the Gulf of Mexico, and has 13 feet of water in the channel. The Rio Grande is navigable for over 400 miles; the Red River, Neuces, Angelina, Trinity, and some other streams are navigable during the season for considerable distances. The Canadian River, in the north, and the Brazos, Colorado, Guadalupe, and San Antonio are among the best-known streams.

Climate.—The climate of Texas shows considerable variation, ranging from the temperate to the semi-tropical, but in general it is remarkably salubrious. The mean annual temperature in the highlands of the extreme northwest is about 56°, in the central division, 65° to 66°, and in the southwest, 72°; and the range of the thermometer is from 35° to 95° Fahr. The rainfall is greatest along the coast and in the south. The aver-

---

and for the summer of 79° Fahr. The rainfall varies from 28 to 32 inches. The prevailing winds are from the Gulf of Mexico, and "northers" (cold wind and rain-storms from the north) are of seldom occurrence. The climate is a happy medium between the humidity and heat of the coast region and the highly rarcified and cold air of the Rocky Mountains. The atmosphere is balmy, and at times invigorating. Combining, as it does, the stimulating effects of mountain air with the mildness of a semi-tropical climate, consumptives, under the writer's personal observation, have frequently recovered appetite and strength through its influence.

age precipitation at Austin for a series of years was found to be 34·54 inches, at Fort Belknap, in Young County, about 22 inches, and in the northwest from 12 to 16 inches.

Population, 1,591,749.

Inhabitants to the square mile, 5·9.

Deaths per 1,000 Inhabitants.

The State	1·0	Grayson	1·6
Group 1	1·3	Grimes	1·5
Cameron	2·8	Guadalupe	0·7
Galveston	1·4	Harrison	1·9
Harris	1·6	Hill	0·7
Remainder of group	0·9	Hopkins	1·2
Group 2	0·9	Houston	0·4
Anderson	1·2	Hunt	1·2
Austin	0·6	Johnson	0·9
Bastrop	0·6	Kaufman	1·1
Bell	0·7	Lamar	1·2
Bexar	1·5	Lavaca	0·8
Bosque	0·8	Leon	0·2
Bowie	1·4	Limestone	1·1
Brazos	0·9	McLennan	1·1
Caldwell	1·1	Marion	1·4
Cass	1·6	Milam	0·6
Cherokee	1·4	Montague	0·8
Collin	0·8	Montgomery	0·7
Colorado	0·7	Nacogdoches	0·9
Cooke	1·2	Navarro	0·8
Coryell	0·9	Panola	0·8
Dallas	1·6	Parker	0·9
Denton	0·7	Red River	0·4
De Witt	0·4	Robertson	0·6
Ellis	1·3	Rusk	1·0
Erath	0·6	Smith	0·9
Falls	0·7	Tarrant	1·0
Fannin	1·0	Travis	1·5
Fayette	1·0	Upshur	1·7
Freestone	0·2	Van Zandt	1·2
Gonzales	1·2	Walker	0·8

Deaths per 1,000 Inhabitants.

Washington	0·5	Wood	1·1
Williamson	0·8	Remainder of group	0·7
Wise	0·4	*Group 3*	0·8

State of Kansas.

Topography.—Kansas has an extreme length east and west of 410 miles; a breadth of about 210 miles; and an area of 82,080 square miles, or 52,531,200 acres. The general surface is an undulating plateau, with a gentle slope from the western border to the Missouri. The extreme elevation reached is 3,800 feet, while at the mouth of the Kansas River the land lies 750 feet above the level of the sea. The average altitude is about 2,375 feet. There are no mountains in Kansas, but the scenery is redeemed from monotony by the rich grass-covered hills and the fertile river-valleys, while the Arkansas and Republican Rivers are bordered by bold bluffs from 200 to 300 feet in height. The Missouri furnishes a water frontage of 150 miles on the east, and near the Missouri State line receives the Kansas, which is formed by the confluence of the Republican and Smoky Hill Rivers near Junction City, and intersects the State throughout its entire length. The Smoky Hill River rises near the Rocky Mountains, in Colorado, and receives in Kansas the Saline and Solomon Rivers, each over 200 miles long. The Republican River rises in southern Colorado, flows through northwestern Kansas into Nebraska, and turning southeast joins the Kansas. The Arkansas River has its sources in the Rocky Mountains in Colorado, and passes through Kansas in an easterly and southeasterly direction, having nearly 500 miles of its windings within this State. The Osage River rises in the east, and, after a southeast course of 130 miles, enters Missouri; while the Neosho has its source in the central part of the State, and after a southeast course of 200 miles, during which it receives the Cottonwood and other important streams, passes into the Indian Territory. Few of the rivers are navigable, but nearly all furnish abundant water-power.

Climate.—The winters of Kansas are comparatively mild, the summers warm but not oppressive, and the atmosphere extraordinarily pure and clear at all seasons. The mean annual temperature is about 53° Fahr.; spring, 52°; summer, 76°; autumn, 54°; winter, 29°. The highest temperature recorded is 100° Fahr., and the lowest −6°, these extremes having been only reached on a very few occasions since the settlement of the country. The rainfall averages from 31 to 45 inches per annum, the greatest precipitation being in the eastern division. The western part of Kansas consists of an extensive plateau, which has a general elevation of from 1,500 to 4,000 feet. The mean annual temperature is 49° Fahr., with a moderate range throughout the year, notwithstanding its considerable diurnal changes. Kansas is a very healthy State, entirely free from miasmatic diseases, and highly favorable to consumptives and those suffering from bronchial or pulmonary complaints, to whom the pure, free atmosphere seldom fails to afford relief.

Population, 996,096.

Inhabitants to the square mile, 12.

Deaths per 1,000 Inhabitants.

The State	1·1
Group 1	1·1
Allen	1·9
Atchison	1·3
Bourbon	1·1
Brown	1·3
Butler	0·5
Chautauqua	0·5
Cherokee	1·0
Clay	1·2
Cloud	0·6
Coffey	0·9
Cowley	1·2
Crawford	0·8
Dickinson	1·7
Doniphan	0·9
Douglas	1·1
Elk	0·7
Franklin	1·4
Greenwood	0·9
Harvey	1·3
Jackson	0·8
Jefferson	1·6
Jewell	1·2
Johnson	1·4
Labette	0·8
Leavenworth	1·2
Linn	1·3
Lyon	0·9
McPherson	1·1
Marion	0·4
Marshall	0·9
Miami	1·7
Mitchell	0·6
Montgomery	1·3
Nemaha	1·2
Neosho	1·3
Osage	1·0
Ottawa	1·0
Pottawatomie	1·4
Reno	1·6
Republic	0·6
Riley	1·8
Saline	0·7
Sedgwick	0·9
Shawnee	1·9
Sumner	0·5
Washington	1·0
Wilson	1·2
Wyandotte	1·8
Remainder of group	1·0
Group 2	1·0
Barton	1·5
Osborne	0·8
Phillips	0·9
Smith	1·2
Remainder of group	0·9

State of Nebraska.

Topography.—The surface of Nebraska constitutes a vast plain, with undulating prairies of great extent, diversified by a few low hills or ridges, and without mountains of any size, except in the extreme west and northwest, where the lower slopes of the Rocky Mountains and the broken country of the Black Hills begin. From the west and northwest the land slopes gradually to the

Missouri River, which washes the eastern and northeastern borders of the State. The drainage is toward the Missouri by the Platte River and its tributaries, the Niobrara, and the Republican and Blue Rivers, which extend into Kansas. The valley of the Platte, which stretches across the center of the State from west to east, and the whole southern portion of Nebraska, are extremely fertile and well watered. The western half is best adapted for grazing purposes, being a constant succession of natural pastures. About 30,000 square miles of the eastern division consist of bottom and prairie lands of exuberant fertility. Nebraska has a width from north to south of about 210 miles; its greatest length in the central part is about 420 miles; area, 76,855 square miles, or 49,187,200 acres.

Climate.—Nebraska might with propriety be termed a highland State, forming as it does a part of the great interior slope which extends from the base of the Rocky Mountains to the Missouri River. Over the wide prairies the mountain-breezes sweep at will, and, owing to the splendid drainage facilities, the dry, exhilarating atmosphere is untainted by any malaria. The mean annual temperature at Omaha is about 48° Fahr.; winter mean, 22°; and summer, 70°. The total snow- and rain-fall east of the 100th meridian is 26 inches, the greatest amount of rainfall being in May and June. In the west and southwest it is much less, and in some places not more than 17 to 19 inches annually.

Population, 452,402.

Inhabitants to the square mile, 5·8.

Deaths per 1,000 Inhabitants.

The State	0·9	York	0·9
Group 1	0·6	Remainder of group	0·9
Adams	0·6	*Group 2*	1·0
Clay	0·3	Cass	0·4
Dodge	0·4	Douglas	0·7
Fillmore	0·3	Nemaha	0·7
Gage	0·9	Otoe	1·7
Lancaster	0·9	Richardson	1·1
Saline	1·0	Remainder of group	1·1
Saunders	0·9	*Group 3*	0·8
Seward	0·4		

State of Minnesota.

Topography.—Minnesota occupies nearly the center of the Continent of North America. The surface of the State is an undulating plain, with an average elevation of 1,000 feet above the sea, but in the northeast there is a group of low sand-hills known as the "Hauteurs des Terres," or "Heights of Land," which rise about 600 feet higher. Its extreme length north and south is 380 miles, and its breadth varies from 183 miles in the middle to 262 miles on the southern and 337 near the northern line; the total area being 83,365 square miles, or 53,353,600 acres. There are over 7,000 small lakes in the State, varying from 1 to 30 miles in diameter, while several of them have an area of from 100 to 400 square miles. The Mississippi rises in Lake Itasca, and flows for nearly 800 miles through the State, receiving the Minnesota at Fort Snelling, 5 miles above St. Paul. The Red River of the North rises in Elbow Lake, turns southwest and north, and empties into Lake Winnipeg. The St. Louis River, which rises in the northeast, falls into Lake Superior, and forms the first link in the chain of rivers and lakes of the St. Lawrence system. There is much really beautiful scenery in Minnesota, and, although it is destitute of mountains, the limestone cliffs of the upper Mississippi, and the perpendicular walls of rock between which the St. Croix forces its way, are very picturesque. The celebrated Falls of St. Anthony, at Minneapolis, are the best known of the many cataracts to be found in this State.

Climate.—The salubrity of the climate of Minnesota is well known, and the purity of the air and dryness of the winters render the State a chosen place of recuperation for those suffering from pulmonary complaints. The summers are warm, with breezy nights, and two thirds of the total rainfall of 35·50 inches occurs during the months of June, July, August, and September. The winters are cold, clear, and dry, and the snowfall generally light. The range of temperature is considerable—the summer mean being 70·50°; winter mean, 25°; and the annual mean at St. Paul, 47° Fahr.

Population, 780,773.

Inhabitants to the square mile, 9.

Deaths per 1,000 Inhabitants.

The State	1·0	Wabasha	0·7
Group 1	1·1	Washington	0·9
Dakota	1·6	Winona	1·2
Goodhue	0·9	Wright	0·7
Hennepin	0·5	Remainder of group	0·7
Minneapolis (city)	1·7	*Group 2*	1·0
Houston	1·3	Blue Earth	0·7
Saint Paul (city)	1·1	Brown	1·0
Stearns	1·3	Carver	1·0

Deaths per 1,000 Inhabitants.

Dodge	1·8	Renville	0·7
Faribault	1·2	Rice	1·2
Fillmore	0·8	Scott	1·0
Freeborn	1·7	Sibley	0·6
Kandiyohi	0·8	Steele	1·2
Le Sueur	0·3	Waseca	1·2
McLeod	1·1	Remainder of group	0·9
Meeker	0·5	*Group 3*	1·1
Mower	0·8	Otter Tail	0·9
Nicollet	1·1	Polk	0·3
Olmsted	1·0	Remainder of group	1·6

Territory of Dakota.

Topography.—Dakota has an average extent north and south of 450 miles, a breadth of 350 miles, with an area of 149,100 square miles, or 95,424,000 acres. There are still 27,550 Indians in the Territory, seven eighths of the whole number being Sioux. These are divided into numerous bands, and are mostly on reservations west of the Missouri and north of the Nebraska frontier. The amount of land held by them is 41,999,456 acres, of which over 5,000,000 acres are tillable; only 10,500 acres, however, are under cultivation. The Territory forms a vast elevated plateau, crossed by several minor ranges of hills, which in the southwest almost deserve the name of mountains. The general elevation of the country is from 1,000 to 2,500 feet above the sea, and the highest peaks of the Black Hills are nearly 7,000 feet above sea-level. The Missouri River crosses Dakota from the northwest to the southeast corner, and is navigable throughout the Territory. It receives the Yellowstone on the Montana border, in latitude 48° north, and has also as tributaries the Little Missouri, White, Big Cheyenne, and Niobrara Rivers, the latter having most of its course in Nebraska. The Dakota rises in Devil's Lake, in the northeast, and has a length of 400 miles to its junction with the Missouri near Yankton. The Red River of the North, flowing north into British America, forms the eastern boundary for about 250 miles, and receives no less than eight considerable streams from Dakota. The Vermillion and Big Sioux in the southeast are each more than 150 miles long. There are a large number of lakes and ponds, mostly in the east and north. Devil's Lake, the water of which is brackish, is 40 miles long and from 4 to 12 miles wide. Other important lakes are Albert, Poinsett, Tchanchicaha, Traverse, Whitewood, and Big Stone.

Climate.—The temperature varies during the year from 20° below zero to 100° Fahr.; July and August being the warmest months, and December, January, and February the coldest. In the north the winters are severe and much snow falls, but the climate of the south is mild and pleasant. The atmosphere is clear and dry, and, owing to the elevation, malarial diseases are unknown, while pulmonary complaints are rare. Spring opens earlier than in the Eastern States in the same latitude. The annual rainfall averages 20 inches.

Population, 135,177.

Inhabitants to the square mile, 0·9.

Deaths per 1,000 Inhabitants.

The Territory	0·8	*Group 3*	0·2
Group 1	0·9	Lawrence	0·3
Group 2	0·8	Remainder of group	0·0

Territory of Montana.

Topography.—The length of the Territory from east to west varies from 460 to 540 miles; its average breadth is 275 miles; and its area is 146,080 square miles, or 93,491,200 acres, of which 80,651,676 are still unsurveyed. The eastern division embraces the great plains or rolling table-lands, which cover three fifths of the area of the Territory; the Rocky Mountains, with other ranges, occupying the west. The Rocky Mountains form the southwest boundary, from the west line of Wyoming to the intersection of the parallel 45° 40′ north latitude with the 114th meridian of longitude; thence run east for some distance, and from the 112th meridian continue in a northwestern direction to the British possessions. The Bitter-Root Mountains branch off at the eastern extension of the Rockies, and form the western boundary of the Territory for a considerable distance. Other important ranges are the Snow Mountains in the south, and the Belt, Highwood, Judith, and Little Rocky Mountains. The peaks are from 6,000 to 12,000 feet above the sea-level, and the valleys average about 4,000 feet, the mountain-belt having an average breadth of 180 miles. The plains slope gradually toward the east, having an elevation of about 4,000 feet at the base of the mountains, and of 2,000 feet at the Dakota line. The Rocky Mountains form the water-shed, and give rise to the Jefferson, Madison, and Gallatin Rivers, which unite near Gallatin City and form the Missouri. The latter runs north, northeast, and finally east; and the Yellowstone, which has its source in the National Park, in Wyoming, runs north and northeast through Montana, and joins the Missouri near its eastern boundary. Other, important streams are the Flathead, Missoula, Big Blackfoot, Bitter Root, and Milk Rivers. The largest body of water is Flathead Lake, which is about 30 miles long by 10 miles wide, and there are several smaller lakes in the northwest. Timber is abundant on the mountain-slopes, and consists of pine, cedar, fir, and hemlock, estimated to cover in all over 25,000,000 acres. Cottonwood, willow, and alder are found along the streams, and in this respect Montana is much better off than many of her immediate neighbors.

Climate.—The climate of Montana is warmer than that of the Eastern States in the same latitude, and is very dry, the rainfall seldom exceeding 12 inches per annum. In the east the mean annual temperature varies from 41° to 49° Fahr., but in the mountainous region it is lower. The snowfall is heavy in the mountains but light in the valleys, and the climate is favorable for outdoor and agricultural occupations, and the raising of stock.

Population, 39,159.

Inhabitants to the square mile, 0·2.

Deaths per 1,000 Inhabitants.

The Territory	0·4

TERRITORY OF IDAHO.

Topography.—Idaho has an irregular shape. It is 485 miles in length north and south, on the western boundary, and 140 miles on the Wyoming border; 45 miles wide in the north, and nearly 300 miles in the south; and contains, as now constituted, 84,800 square miles, or 54,272,000 acres, of which 47,739,368 are still unsurveyed. The surface is an elevated table-land, from 2,000 to 5,000 feet above the sea-level, with many deep river-valleys, and crossed by numerous mountain-ranges or spurs of the Rocky and Bitter-Root Mountain chains. Many of the peaks are of considerable height, and rise above the snow-line. The most important are the Kootenay Mountains, in the extreme north; the Cœur d'Alene range, south of these; the Salmon and Clear-water Mountains, along the rivers of the same names, and the successive ranges on the head-waters of the Snake River. In the southeast are the Bear River Mountains, and in the south the Three Buttes. Of the total area, about 4,480,000 acres are suitable for agriculture, and 5,000,000 for grazing. One third of the entire area is sterile, and yields nothing but sage-brush and a little buffalo-grass, but it is believed that part of this section can be reclaimed by irrigation. There are 8,000,000 acres of timber and as much of mineral land, while numerous lakes occupy an area of 200,000 acres. The lower slopes of the mountains are covered with extensive pine and cedar forests, and there is much timber in the north. Wheat, oats, barley, and rye flourish in the valleys, and wherever irrigation is possible, but the climate is not suited to corn. The Boisé Valley, which is 55 miles in length by 3 in width, and sheltered by the Boisé Mountains, is the chief agricultural region. The Territory is drained by the Snake River and its tributaries, the Bruneau, Boisé, Weiser, Salmon, Clearwater, Payette, and other smaller streams. The Snake, or Shoshone, River rises in the Yellowstone Park, in western Wyoming, describes an immense curve through southern Idaho, and forms the western boundary of the Territory for about 200 miles, after which it turns west into Washington Territory, where it joins the Columbia. It is navigable for a considerable distance within and upon the borders of Idaho for light-draught vessels. The American and Shoshone Falls, and the rapids above the latter, are considered scarcely inferior to the falls of the Yellowstone, the Shoshone having a perpendicular descent of over 200 feet.

Climate.—The winters on the mountains are severe, and much snow falls. On the plains the winter temperature is about the same as that of Wisconsin or northern Iowa. In the valleys the climate is milder, with much less snow, and the springs and summers are pleasant, and never oppressively hot. Idaho belongs to the dry region, or dry plains. The annual rainfall scarcely equals one fourth that of the Atlantic States, although a considerable precipitation occurs on the Bitter Root and Rocky Mountains, but in the north and west and in the lower valleys the rainfall is much less, and irrigation is a prerequisite to successful agriculture.

Population, 32,610.

Inhabitants to the square mile, 0·3.

Deaths per 1,000 Inhabitants.

The Territory..	0·6

State of Colorado.*

Topography.—Colorado has an average length east and west of 380 miles, a breadth of 280 miles, and an area of 103,925 square miles, or 66,512,000 acres, divided into thirty-nine counties. There are still unsurveyed 40,657,679 acres. It consists of three natural divisions—the mountain-range, the foot-hills, and the plains. The Rocky Mountains run north and south through the center of the State, and consist of three parallel ranges, with many peaks over 13,000 feet high. Within the space inclosed by these immense mountains are the "Parks," which constitute the most remarkable natural feature of Colorado. These consist of extensive plateaus at an elevation of 9,000 to 10,000 feet above the sea-level, hemmed in on all sides by the mountains, and containing some valuable agricultural land. The plains occupy the eastern part of the State, and comprise about one third of its area. The great "divide" traverses Colorado northwest and southeast, separates the waters of the South Platte and the Arkansas, and forms the water-shed of many of their tributaries. Colorado has numerous streams, the principal ones being the North and South Platte, and the Arkansas, Snake, White, and Green Rivers, most of which flow

---

* Valuable data regarding the principal health resort for consumptives, in this State, may be found in a paper, "Comparing Colorado Springs with Davos-Plaz, as Winter Health Resorts," by Clinton Wagner, M. D., which was read before the "New York Academy of Medicine," October 20, 1887, and published in the "New York Medical Journal" for October 29, 1887.

through rocky cañons and are not navigable. The South Platte has a fall of 6,000 feet between Montgomery and Denver, and one of the cañons of the Arkansas is 1,500 feet in depth. The Rio Grande del Norte rises in the Saguache range of mountains and flows south through the San Luis Park; but the Colorado River can scarcely be considered as belonging to the State from which it derives several of its principal tributaries, and to which it gave a name.

Climate.—The air is drier and the range of temperature less than in the Eastern States in the same latitude. The winters are mild, the summers cool and bracing, and the mean annual temperature is about 49° Fahr. The rainfall ranges from 15 to 20 inches annually, and most of it occurs between May and July. On the mountains the winters are, as a rule, severe, with heavy falls of snow in November and December; but on the plains and in the valleys the mildness and purity of the atmosphere are such as to render Colorado the paradise of invalids, thousands of whom resort there. An abundance of sunshine is one of the important features of this climate, over three hundred clear days in a single year having been noted for certain localities in this State. Heavy wind-storms are common, but cloudy and foggy weather is unknown, and snow seldom remains more than twenty-four hours on the ground except upon the mountains, many of which reach above the snow-line.

Population, 194,327.

Inhabitants to the square mile, 1·8.

*Deaths per 1,000 Inhabitants.**

The State	1·1	*Group 2*	0·5
Group 1	1·9	Lake	0·1
Denver (city)	1·7	Remainder of group	0·6
Remainder of group	2·1		

Territory of Wyoming.

Topography.—The surface is elevated and mountainous, the main chain of the Rocky Mountains extending across the Territory from southeast to northwest, and forming what is known as "the divide." The principal ranges are the Wind River, Big Horn, Laramie, Bishop, and Medicine-Bow Mountains. The Black Hills lie partly in this Territory and partly in Dakota. The Big Horn, Tongue, and Powder Rivers flow north and join the Yellowstone in Montana; the Green River drains the southwest and the Little Missouri the northeast, while the North Platte, rising in Colorado, receives the Medicine-Bow, Laramie, and Sweetwater Rivers in Wyoming and enters Nebraska from the southeast, where there are some smaller streams of little importance. The most interesting of the natural features of Wyoming, and those which have most attracted the attention of travelers, are found in the extreme northwest corner of the Territory, in the section known as the Yellowstone National Park. This wonderful park has a length of 65 miles north and south by 55 miles in width, and an area of 3,575 square miles. No part of it is less than 6,000 feet

* Almost all the deaths reported as occurring from consumption in this State were of persons who had contracted the disease elsewhere.

above the sea, and the snow-covered mountains that hem in the valleys on every side rise to a height of 12,000 feet. It is a land of wonders, with its grand cañons and geysers, its beautiful lakes and rivers, with cataracts, cascades, and rapids of unexampled beauty, and mountains towering far above the deep and rugged valleys through which the rapid streams flow. The geysers, or boiling springs, are situated near the Firehole River, the Middle Fork of the Madison, which forms one of the three principal sources of the Missouri. There are several hundred springs, of which the Beehive, Giantess, Old Faithful, the Turban, the Giant, and the Grand Geyser are the largest. Wyoming is situated between latitude 41° and 45° north and longitude 104° and 111° west. It has a length east and west of about 350 miles and a breadth of about 275 miles, and forms an almost perfect quadrangle, with an area of 97,890 square miles, or 62,649,600 acres, of which 9,079,186 are surveyed into sections and 42,638 are improved.

Climate.—The climate is severe in the mountainous regions, but mild and salubrious in the sheltered valleys. The air is pure and bracing and the rainfall light, not exceeding 15 inches per annum, and in some parts even less. The mean temperature at Cheyenne (6,058 feet above the sea) in July, the warmest month, is about 71°; in January, the coldest, 12°; and the mean for the year not lower than 43·6° Fahr. A maximum of 98° is recorded in one year, and a minimum of 38° Fahr. The soil of the valleys is a fertile loam, but irrigation is needed for the successful prosecution of agriculture.

Population, 20,789.

Inhabitants to the square mile, 0·2.

Deaths per 1,000 Inhabitants.

The Territory	0·2

TERRITORY OF ARIZONA.

Topography.—The area of the Territory of Arizona is 113,020 square miles, or 72,332,800 acres, of which 67,098,366 are unsurveyed. The middle and northeastern portions of the Territory consist of plateaus which have an elevation of from 3,000 to 8,000 feet above the sea, and are here and there dotted by volcanic cones rising 2,500 feet above the plateaus. The southern portion is a plain with a slight elevation above the sea, amounting to only 200 feet at the mouth of the Gila. The mountain-ranges, of which there are many, have generally a northwest and southeast course, with the exception of the Mogollon range in the east, which runs nearly east and west, joining the Sierra Blanca. The Sierra Prieta and the Aztec range, in central Arizona, are flanked by foot-hills, which sink gradually to the level of the table-land on the northeast, and of the *mesas* sloping toward the Colorado River in the southwest. The highest mountain is the San Francisco, a volcanic cone, whose summit is 11,000 feet above the sea. The Colorado, which is the largest and the only navigable river, is formed by the junction, in southern Utah, of the Green and Grand Rivers, and flows southerly along the western boundary of Arizona, emptying into the Gulf of California just south of the southern line of the

Territory. This river has during the course of centuries cut for itself a deep channel through the rocks, so that for long distances it flows between perpendicular walls 7,000 feet in height. It is navigable for a distance of 612 miles from its mouth; above that point it becomes shallow, except in the rainy season, has a very swift current, and is filled with rapids. Its principal tributaries are the Gila, which has its source in New Mexico, and flows in a southwesterly course until it joins the Colorado, about 180 miles above the Gulf of California; the Colorado Chiquito, which rises in the northwest; and the Bill Williams Fork. Here, as in New Mexico, agriculture can be carried on only where irrigation is practicable, which applies to about five per cent of the total area. There are desert tracts covered with shifting sands, which are utterly unfit for cultivation, and much of the Territory south of the Gila is an arid waste. But the soil in the river-bottoms and in the mountain-valleys of middle and eastern Arizona is of great richness. Pine and cedar grow on the mountains in the central and northern part of the Territory, and walnut, cherry, and cottonwood are found along the streams. On the plains south of the Gila only the cactus, artemisia, and mesquite can live.

Climate.—The climate is mild and generally healthful, lung and malarious diseases being almost unknown. The summer temperature of the treeless plains in the south is intensely hot, the thermometer often indicating 118° Fahr., and rarely falling in winter below 34°. In the central and more elevated portion of the Territory the temperature is moderate, seldom exceeding 90° in summer. Snow falls on the mountains, but remains only a short time. The rainfall along the Gila averages from 4 to 5 inches, while at the base of the range it rises to 25 or 30. Showers are most frequent in July and August.

Population, 40,440.

Inhabitants to the square mile, 0·3.

Deaths per 1,000 Inhabitants.

The Territory (forms one group).	0·4	Remainder of group.	0·6
Pima	0·2		

Territory of New Mexico.

Topography.—New Mexico has a length on the eastern boundary of 345 miles, and on the western of 390, with an average breadth, north of the thirty-second parallel, of 335 miles. Its area is 122,580 square miles, or 78,451,200 acres, of which 67,024,990 are unsurveyed. The region now known as Arizona, obtained from Mexico by the Gadsden Treaty of 1853, was annexed to New Mexico the following year, and formed a part of the Territory until 1863. In 1861 a tract of 14,000 square miles, lying east of the Rocky Mountains, between the thirty-seventh and thirty-eighth parallels, was annexed to Colorado. New Mexico as now constituted consists of a number of high, level plateaus, intersected by mountain-ranges, often rising into high peaks, between which lie fertile valleys. The Rocky Mountains, before entering the Territory, divide into two ranges: the one on the east, the loftier of the two, ending near Santa Fé; and the other, known as the Sierre Madre, of lower elevation, and with numerous passes, extending to the southward until it reaches the Sierra Madre of Mexico. Almost two thirds of the Territory is east of this range. The region to the west, which has not been thoroughly explored, consists of high table-lands and isolated peaks. East of the eastern range the land slopes gradually to the Mississippi. The Staked Plain, an elevated region, unwatered and without wood, extends into the south-eastern part of the Territory. The principal river is the Rio Grande del Norte, which, rising in Colorado, flows south through New Mexico, and, continuing on its course toward the Gulf, forms the boundary between Texas and Mexico. Its principal affluent is the Pecos, which, rising in the eastern part of the Territory, empties into the Rio Grande in Texas. In the northeast rises the Canadian, which empties into the Arkansas; and in the southwest the Gila, which flows into the Gulf of California. The valley of the Rio Grande has an elevation of 3,000 feet above the sea near the southern boundary, and of nearly 6,000 feet at the point where it crosses the Colorado line. On each side of this river, which is not navigable, the mountain-ranges rise to an altitude of from 6,000 to 12,000 feet above the sea, the summits of the loftier peaks being above the snow-line. Timber is not abundant. The mountains are covered with pine, spruce, and fir; nut-pine and cedar grow on the foot-hills, and sycamore and cottonwood in the river-valleys.

Climate.—Owing to the differences in elevation, the climate varies greatly. The mean temperature at Santa Fé, with an elevation of 6,862 feet, is: spring, 49·70°

Fahr.; summer, 70·4°; autumn, 50·6°; winter, 31·6°; year, 50·6°. The thermometer rarely rises above 88°, or sinks below −5°. Pulmonary complaints are infrequent, but, owing to the rarity of the atmosphere, pneumonia and similar complaints are frequent. The rainfall is very slight, sometimes not exceeding ten inches per year.

Population, 119,565.

Inhabitants to the square mile, 0·9.

Deaths per 1,000 Inhabitants.

The Territory	0·4	Rio Arriba	0·5
Group 1	0·1	Santa Fé	0·7
San Miguel	0·2	Taos	0·9
Remainder of group	0·0	Valencia	0·2
Group 2	0·5	Remainder of group	0·4
Bernalillo	0·2		

State of California.

Topography.—California, the largest State in the Union with the exception of Texas, has an extreme length of 770 miles, an extreme breadth of 330 miles, and an estimated area of 158,360 square miles, or 101,350,400 acres. The Sierra Nevadas and the Coast Range of mountains run northwest and southeast, generally parallel, and are connected in the north and south by transverse ranges. Between the two ranges lie the San Joaquin and Sacramento Valleys. The Yosemite Valley, situated in the midst of the Sierras, forms one of the chief attractions of the State. The Sierra Nevadas have a general elevation of from 8,000 to 15,000 feet. In the southern part of the main range is Mount Whitney,

15,000 feet high. In the north Mount Shasta, a bare volcanic peak of 14,400 feet in height, is the best known. The Coast Range is inferior in grandeur to the Sierras, having an average elevation of 2,500 to 4,000 feet. The Sacramento River rises near Mount Shasta, and flows south until in latitude 38° it unites with the San Joaquin. The latter has its origin in Tulare Lake, and its course is northerly until it joins the Sacramento. After receiving the San Joaquin the Sacramento flows west to the sea. The Klamath has its origin in Oregon and flows through the northwest part of California, and the Colorado forms in part the southeast boundary and empties into the Gulf of California. The principal lakes are Tulare and Mono. Lake Tahoe forms part of the boundary between California and Nevada. The principal bay is that of San Francisco, which is forty miles long and nine wide, and forms the best harbor on the western coast of North America.

Climate.—The variation in climate, owing to the difference in elevation and latitude, is great. On the coast the winters are mild and the summers extremely pleasant. At San Francisco the summer mean is 60° Fahr., that of winter 51°, and of the year 56°. In the interior the summers are much warmer, and in the Sacramento Valley the mercury often reaches 100°. In the twenty-three years, 1850–1872 inclusive, the rainfall at the same city varied from 7 to 50 inches per annum, and extreme variability from year to year is shown in other parts of the State. In the south the average is not over 10 inches, and at Fort Yuma even less. The heavy snows which rest on the Sierras partially correct the irregularity of the rainfall.

Population, 864,694.

Inhabitants to the square mile, 5·4.

Deaths per 1,000 Inhabitants.

The State*	2·0	*Group 2*	2·3
Group 1	1·6	Alameda	1·4
Amador	1·4	Oakland (city)	1·7
Butte	1·0	Contra Costa	2·0
Colusa	1·7	Humboldt	1·6
El Dorado	1·0	Los Angeles	2·7
Napa	2·6	Marin	1·5
Nevada	0·6	Mendocino	1·2
Placer	1·3	Monterey	0·9
Sacramento	2·1	San Francisco (city)	3·0
San Joaquin	3·2	Santa Clara	2·0
Tulare	1·7	Santa Cruz	1·8
Yolo	1·4	Solano	0·6
Yuba	3·5	Sonoma	1·7
Remainder of group	1·2	Remainder of group	1·8

State of Nevada.

Topography.—Nevada has an extreme length north and south of 485 miles; its greatest breadth through the center is about 320 miles; area, 110,700 square miles, or 70,848,000 acres, with 58,436,498 still unsurveyed. The surface is an elevated table-land, with an average altitude of 4,500 feet above the ocean, and broken by parallel ranges of mountains running from north to south, which attain a height of from 1,000 to 8,000 feet. The Sierra Nevadas, which reach an elevation varying from 7,000 to 13,000 feet, form a part of the western

* Many of the deaths from consumption indicated for this State were of persons who had contracted the disease elsewhere.

boundary. It would be difficult to say what part of the State is the water-shed, for the rivers, which are not navigable, run in all directions, and with few exceptions fail to reach the sea. Some empty into lakes or sloughs and others sink into the earth. The Colorado River forms a part of the eastern and southeastern border, and the longest stream is the Humboldt, which rises in the northeastern part of the State and has a course of 300 miles within it, terminating in Humboldt Lake. Lake Tahoe, among the mountains on the California border, is twenty-one miles long and ten miles wide, and has a depth of 1,500 feet. It is more than 6,000 feet above the ocean, but keeps a temperature of about 57° Fahr. the year round. Pyramid and Walker Lakes are also extensive bodies of water and of great depth. The other lakes are little else than marshes formed by the overflow of the streams, and in many cases their waters are alkaline or brackish. Among the most noticeable natural features are the "mud-lakes" and warm springs. Some of the former cover 100 square miles, and are composed of thick alkaline deposits in the dry season, or of a foot or two of very muddy water during the rains. Most of the springs contain sulphur or other mineral ingredients and possess medicinal qualities.

Climate.—The winters are mild, with little snow except upon the mountains, but in the north the thermometer sometimes falls as low as 15° below zero. In the south and east the weather is much more moderate and frosts are rare, but the summer temperature ranges up to 95° and even 105° Fahr., May and June being the hottest months. The rainfall is light, and occurs principally in the spring, or from January to the end of April. The air is invigorating and bracing, and the climate is considered very healthy.

Population, 62,266.

Inhabitants to the square mile, 0·5.

Deaths per 1,000 Inhabitants.

The State (forms one group)....	0·9	Remainder of group.........	0·9
Storey....................	1·0		

TERRITORY OF WASHINGTON.

Topography.—The greatest length of the Territory east and west is 340 miles; greatest breadth, 240 miles; area 69,180 square miles, or 44,275,200 acres, of which 28,836,985 acres are still unsurveyed. The Cascade Mountains traverse it north and south from British Columbia to Oregon, and divide it into two unequal portions, the eastern section containing about 50,000 and the western nearly 20,000 square miles. The highest peak is Mount Rainier, 14,500 feet, and there are several others little inferior. Between Puget Sound and the Pacific the Coast Range attains considerable prominence and culminates in Mount Olympus, 8,100 feet high. There are also the Blue Mountains in the southeast, which extend into the Territory from Oregon. Eastern Washington is an irregular, broken country, and the chief divisions of the western section are the Columbia and Puget Sound basins and the valley of the Chehalis. The Columbia River enters the Territory from the north, traverses its whole breadth, constitutes almost the entire southern boundary, and with its tributaries drains nearly its whole area. The Snake, Walla Walla, Spokane, Colville, and Clarke's Fork are its principal affluents. It is navigable throughout the Territory, and the Snake is navigable from the Idaho border to its junction with the Columbia. The Territory has a coast-line on the Pacific of about 180 miles, and the deep indentation of Puget Sound, with Admiralty Inlet and Hood's Canal, furnishes many excellent harbors. The scenery of the Columbia River is in many places picturesque and even grand, flowing as it does through rocky mountain-gorges and containing a number of cataracts and rapids. Of these, the chief are the Cascades, where the river breaks through the Cascade Mountain range; the Dalles, 40 miles above; Priest rapids, 179 miles above the Dalles; Buckland rapids, 66 miles farther; and Kettle Falls, 274 miles above; the last having a perpendicular fall of fifteen feet. At Vancouver the river is a mile wide, and so great is the force of the current that it overcomes the effect of the tide, and the water on the bar is rendered drinkable.

Climate.—On the western slope there are but two seasons, the dry and the rainy, the latter commencing late in October and lasting until April. The rainfall is from 70 to 125 inches. The winters are mild, with but little snow or ice, and the summers cool and pleasant, the thermometer in July and August seldom reaching 90° Fahr., while the nights are cool, and there is usually a breeze from the sea. At Steilacoom, Puget Sound, the mean temperature of the year is about 50° Fahr.; summer, 63°; winter, 39°. The rainfall averages about 50 inches per annum. The section east of the mountains possesses a drier climate, and the seasons of spring and autumn are more definitely marked. Washington Territory is extremely healthy, and, from the absence of marshes and the elevation of most of the land, is entirely free from miasma.

Population, 75,116.

Inhabitants to the square mile, 1·0.

Deaths per 1,000 Inhabitants.

The Territory	1·3

State of Oregon.

Topography.—The Cascade Mountains, which cross the State from north to south, dividing Oregon into two unequal parts, known as Eastern and Western Oregon, range from 4,000 to 10,000 feet in height, reaching the region of perpetual snow. The principal peaks are Mount Hood, 11,225 feet; Mount Jefferson, 10,200 feet; the Three Sisters and Diamond Peak, each 9,420 feet; and Mount McLoughlin, 11,000 feet. The Coast Range runs parallel with the ocean, at a distance from it of about 25 miles, the general altitude varying from 1,000 to 4,000 feet. Each of the great ranges throws out spurs, and the eastern division is broken by the Blue Mountains, which run northeast and southwest, and have an average altitude of from 5,000 to 7,000 feet. The valleys are deep and irregular, and in many places the rivers cut their way through romantic cañons of great depth. The principal accessible harbors are the mouths of the Columbia and Rogue Rivers, and Tillamook Bay and Port Orford. The State has an average length east and west of about 360 miles, a breadth of 260 miles, and an area of 96,030 square miles, or 61,459,200 acres. The State has many streams, especially in the western half, but few of them are navigable. The Columbia rises in the Rocky Mountains, in latitude 50° 20′, and is navigable for ships 115 miles from the sea, and for steamers 165 miles. It is a rapid stream, and receives nearly all the rivers of Oregon. The Columbia is 1,300 miles in length, and forms the State boundary for about 300 miles. Its numerous cascades, cañons, narrows, and rapids enhance the beauty of the scenery along this stream. The Rogue River, in southern Oregon, and the Umpqua, which flows through the valley of the same name, both take their rise in the Cascade Mountains and empty into the Pacific. Most of the lakes, of which there are a large number, are situated in Grant and Jackson Counties.

Climate.—The climates of the two divisions differ widely, that of the western half being moist and equable, while the eastern never has an excess of rain, and, though somewhat subject to extremes of temperature, the climate is usually pleasant. The summers of the eastern half are dry, there being little rain and less dew, but the crops do not suffer from drought. The mean temperature at the Dalles is, in spring, 53°; in summer, 70·5°; in autumn, 52°; and in winter, 35·5° Fahr.; and the rainfall does not exceed 18 or 20 inches annually. It is very different in Western Oregon, the annual rainfall at Astoria being 60 inches. Snow and ice are here unknown, but on the mountains and elevated table-lands frosts are frequent, and the higher peaks wear their snowy crowns the year through. The mean annual temperature at Astoria is 52°; that of spring, 51°; summer, 61·5°; autumn, 54°; winter, 42·5° Fahr.

Population, 174,768.

Inhabitants to the square mile, 1·8.

Deaths per 1,000 Inhabitants.

The State	1·2	Linn	0·8
Group 1	1·0	Marion	1·9
Wasco	2·0	Multnomah	1·7
Remainder of group	0·7	Remainder of group	1·2
Group 2	1·3		

Territory of Utah.

Topography.—The average length of Utah north and south is about 350 miles; average breadth, about 260 miles; area, 84,970 square miles, or 54,380,800 acres. The country is rugged and broken, and is separated into two unequal sections by the Wahsatch Mountains, which cross it from northeast to southwest. Extending east from the Wahsatch, along the southern border of Wyoming, are the Uintah Mountains. Other prominent ranges are the Roan, Little Sierra, Lasal, Sierra Abago, San Juan, and Sierra Panoches. In the southeast are extensive elevated plateaus, and in the west a series of disconnected ridges and mountain-ranges, generally extending from north to south. East of the Wahsatch the drainage is mostly by the streams which form the Colorado. Of these the chief are Grand and Green Rivers. White, Uintah, and San Rafael are tributary to Green River.

The Rio Virgin, in the southwest, joins the Colorado in Nevada. Among the lakes, the largest is the Great Salt Lake in the northwest, which is 75 miles long and about 30 broad. Utah Lake is a beautiful sheet of fresh water, having an area of about 130 square miles, and closely hemmed in by mountains. It is connected with the Great Salt Lake by the Jordan River. Bear Lake is on the Idaho border, and partly in that Territory. The Sevier River, rising in the southern part of Utah, flows north for 150 miles, receiving the San Pete and other smaller streams, then bends southwest and forms Sevier Lake, about 100 miles southwest of the Great Salt Lake.

Climate.—The climate for the most part is mild and healthful. The mean annual temperature east of the Wahsatch Mountains is from 38° to 44°, and west of that range from 45° to 52° Fahr., while in the valley of the Rio Virgin and in the southwest generally the summers are dry and hot. The rainfall averages 15 to 16 inches per annum, and sometimes reaches 20 inches in the north. Most of the rain falls between October and April; spring opens in the latter month, and cold weather seldom sets in before the end of November. In the mountainous districts the winters are severe, and the snowfall is heavy.

Population, 143,963.

Inhabitants to the square mile, 1·6.

Deaths per 1,000 Inhabitants.

The Territory (forms one group)	0·4	Utah	0·2
Cache	0·4	Weber	1·0
Salt Lake	0·5	Remainder of group	0·4
Sanpete	0·2		

Table showing Relative Healthfulness of States and Territories.

STATES AND TERRITORIES.	Deaths from all causes per 1,000 inhabitants.	Deaths from consumption per 1,000 inhabitants.
Arizona	7·19	0·4
Montana	8·58	0·4
Wyoming	9·09	0·2
Dakota	9·64	0·8
Idaho	9·90	0·6
Washington	10·05	1·3
Oregon	10·66	1·2
Minnesota	11·57	1·0
Nevada	11·69	0·9
Florida	11·72	0·9
Iowa	11·92	1·1
West Virginia	11·99	1·5
Michigan	12·06	1·5
Wisconsin	12·17	1·2
Mississippi	12·88	1·0
Nebraska	13·10	0·9
Colorado	13·10	1·1
Ohio	13·32	1·8
California	13·32	2·0
Georgia	13·97	1·1
Alabama	14·20	1·3
Kentucky	14·38	2·2
Illinois	14·62	1·4
Connecticut	14·74	2·2
Maine	14·87	2·8
Pennsylvania	14·91	1·8
Delaware	15·08	2·4
Vermont	15·11	2·4
Kansas	15·21	1·1
North Carolina	15·36	1·5
Louisiana	15·44	1·6
Texas	15·53	1·0
Indiana	15·77	1·9
South Carolina	15·79	1·5
New Hampshire	16·09	2·4
Virginia	16·13	1·9
New Jersey	16·33	2·3
Utah	16·76	0·4
Tennessee	16·80	2·4
Missouri	16·88	1·6
Rhode Island	17·00	2·3
New York	17·37	2·5
Maryland	18·09	2·4
Arkansas	18·45	1·1
Massachusetts	18·59	2·9
New Mexico	20·37	0·4
District of Columbia	23·60	4·4

Although the United States possesses almost every variety of climate, it is very difficult accurately to classify its types. This is due in most part to the fact that definite climatic conditions do not prevail to an equal degree, throughout the year, for any given locality. While various combinations of atmospheric tension, temperature, humidity, and precipitation (which always occur in deference to natural laws) distinguish one type of climate from another, they also give rise to the changes which occur from time to time in the climate of any given locality.

The following classification, which comprises most of the climates of this country to which consumptives resort for relief or recovery, commends itself to the writer as sufficiently accurate for general consideration:

1. Climate cool and moderately moist.—General elevation, 2,000 feet. Western slope of the Appalachian chain. Adirondack, Catskill, Alleghany, and Cumberland Mountains.

2. Climate moderately warm and moderately moist.—Western North Carolina: Asheville—elevation, 2,250 feet. Western South Carolina: Aiken. Georgia: Marietta, Thomasville.

3. Climate warm and moist.—Florida (equable). Southern California, coast-region (equable).

4. Climate warm and moderately dry.—Elevation about 2,000 feet. Southwestern Texas, southern California, inland.

5. Climate cool and moderately dry.—Elevation about 1,000 feet. Minnesota, Nebraska, Dakota.

6. Climate cool and dry.—Elevation from 4,000 to 7,000 feet. Montana, Wyoming, Colorado, northern New Mexico, and western Kansas.

7. Climate warm and dry.—Elevation, 3,000 to 5,000 feet. Southern New Mexico and southern Arizona.

V.

CONSUMPTION IN THE U. S. ARMY, AND METEOROLOGICAL REPORTS.

*Ratio of Deaths from Pulmonary Diseases per Mean Strength U. S. Army.**

	Mean Strength.	DISCHARGES PER 1,000 FOR		DEATHS PER 1,000 FROM		LOSS PER 1,000 FROM	
		Consumption.	Diseases of the respiratory organs.	Consumption.	Diseases of the respiratory organs.	Consumption.	Diseases of the respiratory organs.
U.S. Army, white troops, 1870–'74.	25,989	3·828	1·395	1·462	1·462	5·29	2·867
U. S. Army, colored troops, 1870–'74	2,530	2·962	·296	2·47	3·162	5·432	3·458
Arizona							
California							
Colorado							
Dakota							
Kansas							
Minnesota							
Montana	564,646	......	......	1·735	·953	1·735	·953
Nebraska							
New Mexico							
Oregon							
Texas							
Utah							
Washington							
Wyoming							
United States	5,804,616	......	......	2·486	·788	2·486	·788

* Bell's "Climatology," etc., *loc. cit.*

Notes to Table on page 199.

a. Temperature and rainfall estimated for October, 1873.
b. Temperature and rainfall not given.
c. Rainfall estimated for February, 1871, and June, 1872.
d. Rainfall estimated for January, 1874.
e. Temperature and rainfall estimated for the year ending June 30, 1874.
f. Temperature and rainfall for last two years only.
g. Rainfall estimated for January, 1871.
h. Rainfall estimated for January, February, March, 1871.
i. Temperature and rainfall for first two years only.
j. Temperature and rainfall imperfect and incomplete.

Ratio of Diseases and Deaths from Consumption to Total Number of Diseases and Deaths from all Causes at Military Posts of the United States Army, 1870–1874. Abstract of Circular No. 8.

MILITARY POSTS.	Altitude.	Mean annual temperature.	Mean rainfall.	No. of enlisted men.	No. of cases of disease.	No. of cases of consumption.	Per cent of consumption to all other diseases.	Total deaths.	Deaths from consumption.	Per cent of deaths fr'm consumpt'n to deaths from all causes.
Northern Coast.		Deg.	In.							
Ft. Columbus, N. Y.	....	51·58	46·69	2,184	4,618	17	·0036	68	6	·0882
Ft. Adams, R. I.	....	47·95	37·84	1,020	1,514	3	·0019	9	1	·1111
Ft. Independence, Mass.	....	46·46	40·78	227	537	7	·0130	1	0	·0000
Ft. Preble, Me. *a*	....	46·10	30·60	170	72	2	·0279	3	0	·0000
Average	....	50·51	30·01	3,601	6,741	29	·0043	81	7	·0864
Southern and Gulf Coasts.										
Ft. Monroe, Va.	....	58·18	42·16	1,361	1,636	16	·0097	13	3	·2307
Charleston, S. C. *b*	....	......	......	601	903	5	·0055	6	2	·3333
Key West Barracks, Fla. *c*	....	78·09	44·87	372	675	4	·0045	8	1	·1250
Ft. Brown, Tex.	....	72·41	20·80	606	1,832	4	·0030	9	3	·3333
Average	Sea-level.	69·56	35·94	2,940	4,846	29	·0056	36	9	·2500
Northern Interior.	Ft. above sea-level.									
West Point, N. Y. *d*	157	51·24	39·91	1,210	1,627	13	·0080	4	0	·0000
Plattsburg, N. Y.	186	43·92	20·86	283	648	4	·0060	0	0	·0000
Ft. Snelling, Minn.	840	42·98	20·51	434	1,106	3	·0027	3	0	·0000
Omaha Barracks, Neb. *e*	960	50·91	31·58	1,573	2,307	3	·0013	12	1	·0833
Average	....	47·25	30·46	3,500	5,688	23	·0040	19	1	·0526
Southern Interior.										
Columbia, S. C. *f*	300	64·06	53·01	885	1,722	8	·0046	13	2	·1546
Jackson Barracks, La. *b*	10	......	......	975	1,990	6	·0030	15	2	·1333
Ringgold, Tex.	521	74·02	14·75	448	711	7	·0098	10	0	·0000
Ft. Leavenworth, Kan. *g*	?	51·68	57·04	1,492	2,901	6	·0020	16	1	·0625
Average	....	63·02	41·60	3,800	7,333	27	·0036	54	5	·0925
Interior, 1,000 to 2,500 ft.										
Atlanta, Ga.	1,078	62·60	63·46	1,167	1,719	8	·0046	18	3	·166
Ft. Hays, Kan. *h*	1,893	54·25	28·02	814	1,684	9	·0053	12	3	·250
Ft. Sully, Dakota	1,660	47·01	16·39	876	1,302	3	·0023	4	1	·2500
Ft. Larned, Kan.	1,932	52·74	16·24	450	864	1	·0011	7	2	·0285
Camp McDowell, Ariz.	1,800	70·07	11·28	698	957	2	·0020	10	2	·2000
Ft. Sill, Ind. Ter.	1,700	61·06	20·29	1,936	3,058	8	·0026	24	4	·1667
Average	....	57·95	25·78	5,941	9,584	31	·0032	85	15	·1764
Interior, above 2,500 ft.										
Camp Douglas, Utah.	4,904	50·46	16·47	1,290	3,676	3	·0008	16	1	·0625
Ft. Stockton, Tex.	4,950	64·97	12·49	709	628	5	·0078	11	1	·0909
Santa Fé, N. M. *i*	6,850	53·63	14·80	200	217	0	·0000	3	1	·3333
Ft. Bridger, Wyoming	7,010	38·58	10·28	460	676	5	·0074	3	0	·0000
Ft. Ellis, Montana *j*	5,800	......	......	1,043	703	2	·0028	5	0	·0000
Camp Bidwell, Cal.	4,680	49·07	15·05	341	302	2	·0006	6	1	·1666
Average	....	51·34	13·81	4,033	4,202	17	·0040	44	4	·0900

*Monthly and Annual Mean Actual Barometer (corrected for Temperature and Instrumental Error only) deduced from Observations taken at 7 A. M, 3 and 11 P. M., Washington Mean Time, from July, 1880, to June, 1881, inclusive.**

The daily means are obtained by dividing the sum of the 7 A. M., 3 and 11 P. M. observations by three ; the monthly means by dividing the sum of the daily means by the number of days in the month.—(U. S. Signal Service).

STATION.	Elevation of cistern of barometer above m'n sea-level, feet.	1880.						1881.						Annual mean.
		July.	August.	September.	October.	November	December.	January.	February.	March.	April.	May.	June.	
Albany, N. Y.	75	29·877	29·979	29·954	30·015	30·143	29·054	30·070	30·114	29·682	29·707	29·955	29·828	29·947
Alpena, Mich	609	29·312	29·394	29·312	29·852	29·417	29·324	29·403	29·454	29·201	29·328	29·389	29·308	29·352
Atlanta, Ga.	1,131	28·902	28·903	28·932	28·950	29·009	28·904	28·938	28·934	28·722	28·834	28·896	28·823	28·896
Atlantic City, N. J.	13	29·986	30·069	30·051	30·115	30·241	30·019	30·136	30·173	29·735	29·888	30·050	29·902	30·080
Augusta, Ga.	183	29·858	29·890	29·932	29·952	30·044	29·930	29·990	29·980	29·724	29·828	29·852	29·776	29·896
Baltimore, Md.	45	29·957	30·022	30·022	30·086	30·217	30·028	30·124	30·163	29·743	29·889	30·012	29·879	30·012
Barnegat, N. J.	20	29·961	30·044	30·022	30·086	30·218	29·900	30·107	30·147	29·704	29·862	30·025	29·876	30·008
Bismarck, Dak.	1,704	28·155	28·157	28·146	28·187	28·275	28·286	28·277	28·207	28·206	28·181	28·141	28·067	28·190
Boisé City, Idaho	2,768	27·155	27·094	27·250	27·300	27·401	27·153	27·286	27·220	27·222	27·152	27·130	27·071	27·203
Boston, Mass	142	29·812	29·902	29·862	29·924	30·043	29·797	29·920	29·977	29·540	29·661	29·901	29·738	29·840
Brackettville, Tex	1,137	28·808	28·811	28·836	28·922	28·964	28·938	28·940	28·856	28·783	28·766	28·731	28·783	28·841
Breckinridge, Minn.	966	28·931	28·954	28·930	28·952	a 29·086	......	......	......	......	......	......	......	......
Brownsville, Tex.	43	b29·928	29·868	29·919	30·009	30·073	30·057	30·082	c29·998	29·926	29·809	29·858	29·845	29·955
Buffalo, N. Y.	664	29·258	29·323	29·314	29·336	29·416	29·275	29·367	29·404	29·058	29·219	29·315	29·210	29·291
Burlington, Vt.	268	29·700	29·775	29·753	29·801	29·913	29·744	29·874	29·872	29·448	29·542	29·728	29·588	29·728
Cairo, Ill.	377	29·679	29·668	29·730	29·749	29·883	29·800	29·806	29·742	29·542	29·637	29·639	29·581	29·705
Campo, Cal.	2,527	d 27·396	e27·328	27·360	27·415	27·441	27·438	27·482	27·470	27·425	f 27·406	27·348	f 27·358	27·406
Camp Thomas, Ariz.	(†)	g 27·214	27·168	27·212	27·261	27·272	27·283	27·293	27·236	27·217	27·174	27·099	27·116	27·212
Cape Hatteras, N. C.	8	30·019	30·067	30·062	30·094	a 30·192	......	......	......	......	......	......	......	......
Cape Henry, Va.	16	29·984	30·042	30·052	30·109	30·234	30·034	30·133	30·160	29·765	29·918	30·034	29·894	30·029
Cape Lookout, N. C.	15	30·010	30·047	30·054	30·082	30·195	h 30·034	......	......	......	......	......	......	......
Cape May, N. J.	27	29·959	30·036	30·018	30·085	30·214	30·003	30·117	30·157	29·722	i29·870	j30·018	29·873	30·006
Castroville, Tex.	778	29·200	b29·186	29·220	29·307	f 29·297	d 29·282	29·336	k29·246	29·185	i29·182	e29·132	29·137	29·234
Cedar Keys, Fla.	22	30·006	29·988	30·040	30·045	30·140	30·094	30·110	30·100	29·986	30·051	29·978	29·982	30·048
Charleston, S. C.	52	29·968	29·907	30·025	30·066	30·151	c30·048	30·106	30·108	29·812	29·960	29·986	29·910	30·014
Charlotte, N. C	838	29·184	29·222	29·250	29·275	29·352	29·109	29·269	29·276	28·984	29·148	29·103	29·099	29·204

Chattanooga, Tenn.	783	29·256	29·269	29·300	29·396	29·398	29·308	29·339	29·327	30·085	29·203	29·989	30·177	29·267
Cheyenne, Wyo.	6,089	24·141	24·116	24·136	24·105	24·004	23·987	23·010	23·919	23·945	23·996	24·049	24·084	24·068
Chicago, Ill.	661	29·381	29·315	29·328	29·340	29·430	29·364	29·391	29·367	29·155	29·291	29·313	29·284	29·317
Chincoteague, Va.	14	29·988	30·065	30·049	30·110	30·298	30·049	30·139	30·108	29·751	29·908	30·041	29·994	30·081
Cincinnati, Ohio	680	29·393	29·410	29·451	29·466	29·573	29·478	29·494	29·480	29·307	29·348	29·380	29·304	29·416
Cleveland, Ohio	690	29·281	29·329	29·337	29·448	29·442	29·321	29·382	29·369	29·088	29·250	29·315	29·211	29·316
Coleman City, Tex.	1,735(?)	b28·202	c28·204	28·240	28·298	28·345	28·317	28·311	28·343	i28·157	i28·144	28·140	l28·110	28·285
Columbus, Ohio	805	29·178	29·207	29·233	29·248	29·342	29·238	29·204	29·201	28·972	29·135	29·168	29·048	29·194
Concho, Tex	1,888	28·071	28·095	28·136	28·160	28·209	28·181	28·167	28·100	28·040	28·029	28·018	28·003	28·101
Corsicana, Tex.	448	29·564	29·551	29·562	29·654	29·744	m29·711	29·706	29·626	29·547	29·590	29·495	29·492	29·598
Champaign, Ill.	708	......	......	......	n29·262	29·365	29·277	29·292	29·247	29·092	29·108	29·194	29·115	......
Davenport, Ia.	615	29·334	29·353	29·395	29·396	29·531	29·462	29·478	29·431	29·254	29·304	29·351	29·272	29·385
Dayton, Wash.	1,700	......	......	......	......	......	......	o28·410	28·287	28·306	28·274	28·290	28·191	......
Deadwood, Dak.	4,630	25·441	25·410	25·424	25·410	25·372	25·270	25·298	25·292	25·246	25·317	25·372	25·390	25·345
Decatur, Tex.	1,100(?)	d28·844	k28·840	b28·891	f28·948	f28·991	28·950	e28·989	k28·840	28·781	i28·707	l28·750	28·784	28·855
Del. Breakwater, Del.	20	29·978	30·043	30·017	30·002	30·222	30·016	30·028	30·162	29·732	29·887	30·081	29·890	30·016
Denison, Tex.	767	29·297	29·307	29·263	29·305	29·300	29·349	29·335	29·357	29·179	29·174	29·150	29·188	29·247
Denver, Col.	5,294	24·853	24·887	24·678	24·866	24·808	24·710	24·708	24·708	24·717	24·741	24·778	24·733	24·778
Des Moines, Ia.	840	29·111	29·118	29·161	29·105	29·272	e29·244	29·243	29·107	29·067	29·126	29·089	29·007	29·140
Detroit, Mich.	661	29·306	29·361	29·364	29·374	29·461	29·347	29·418	29·405	29·118	29·278	29·332	29·285	29·382
Dodge City, Kan.	2,512	27·409	27·428	27·489	27·458	27·514	27·412	27·434	27·387	27·327	27·358	27·364	27·290	27·408
Dubuque, Ia.	665	29·272	29·298	29·334	29·318	29·423	29·390	29·407	29·377	29·208	29·297	29·294	29·211	29·318
Duluth, Minn.	644	29·268	29·312	29·349	29·287	29·382	29·357	29·410	29·434	29·310	29·315	29·324	29·227	29·328
Eagle Pass, Tex	800	29·143	29·163	29·205	29·300	29·404	29·366	29·375	29·277	29·196	29·109	29·130	29·122	29·237
Eagle Rock, Idaho	4,781	......	......	......	......	......	p25·240	25·268	25·258	25·263	25·233	25·216	25·168	......
Eastport, Me.	61	29·845	29·925	29·909	29·963	30·044	29·706	29·898	29·965	29·605	29·630	29·995	29·784	29·864
El Paso, Tex.	3,770	26·221	26·210	26·357	26·283	26·201	26·201	26·250	26·207	26·180	26·172	26·137	26·151	26·217
Erie, Pa.	681	29·967	29·305	29·308	29·818	29·408	29·282	29·357	29·381	29·065	29·232	29·310	29·202	29·286
Escanaba, Mich.	612	29·304	29·366	29·312	29·314	29·394	29·359	29·417	29·447	29·266	29·340	29·365	29·278	29·344
Florence, Ariz.	1,553	28·351	28·311	28·396	28·451	28·499	28·501	28·535	28·444	28·494	28·396	28·295	28·305	28·400
Ft. Apache, Ariz.	5,004	25·055	25·085	25·070	25·078	25·094	25·026	25·010	25·004	24·984	24·998	24·950	25·000	25·081
Ft. Assiniboine, Mont.	2,650	......	q27·170	27·165	27·206	27·294	27·229	27·221	27·128	27·107	27·127	(r)	(r)	......
Ft. Bennett, Dak.	1,440	......	......	s28·454	28·411	28·539	28·526	28·515	f28·448	28·420	i28·417	28·346	28·268	......
Ft. Benton, Mont.	2,700	27·010	26·998	27·205	b27·254	27·344	27·264	27·267	27·158	27·144	t	u27·161	27·078	......
Ft. Buford, Dak.	1,876	27·915	27·961	27·986	28·029	28·101	28·132	28·112	28·020	28·005	27·922	27·962	27·676	28·002
Ft. Custer, Mont.	3,100(?)	26·840	26·796	26·856	26·889	26·967	26·865	26·869	26·803	26·700	26·802	26·604	26·725	26·833
Ft. Davis, Tex.	4,918	25·200	k25·248	v......	v......	w25·238	25·107	25·157	25·150	25·137	25·155	25·149	25·171	......
Ft. Elliott, Tex.	3,000	27·003	27·282	27·316	27·326	27·352	27·276	27·280	c27·247	x27·225	27·108	27·200	27·155	27·398
Ft. Gibson, Ind. Ter.	510(?)	29·466	29·445	29·495	29·543	29·539	29·599	29·592	29·519	29·404	29·408	29·379	29·345	29·468
Ft. Grant, Ariz.	4,737	25·108	25·184	25·249	25·254	25·219	25·206	25·183	25·109	25·167	25·177	25·167	25·221	25·199
Ft. Griffin, Tex	1,243	28·727	28·744	28·782	28·842	28·887	28·846	28·795	28·660	28·606	28·586	28·568	28·617	28·733
Ft. Keogh, Mont.	2,413	27·252	27·388	27·498	27·594	27·612	27·595	27·558	27·481	27·460	27·444	27·439	27·340	27·466

Monthly and Annual Mean Actual Barometer, July, 1880, to June, 1881, inclusive.—(Continued.)

STATION.	Elevation of cistern of barometer above m'n sea-level, feet.	1880. July.	August.	September.	October.	November	December.	1881. January.	February.	March.	April.	May.	June.	Annual mean.
Ft. McKavett, Tex....	1,217	27·758	27·758	27·796	27·826	27·851	27·830	27·821	27·748	27·696	j27·700	27·684	27·680	27·763
Ft. Missoula, Mont....	3,375(?)	26·744	26·692	26·744	26·814	26·924	26·693	26·769	26·709	26·734	26·690	26·690	26·625	26·737
Ft. St. Michael's, Alaska	....	29·817	29·738	29·582	28·825	29·846	29·870	29·843	30·024	29·808	29·919	29·852	29·818	29·830
Ft. Shaw, Mont.......	3,470(?)	26·267	26·248	26·382	26·430	26·483	26·361	26·383	26·293	26·316	26·337	26·356	26·287	26·345
Ft. Sill, Ind. T........	1,190	28·788	28·775	28·830	28·865	28·941	28·876	28·804	28·798	28·728	28·716	28·699	28·656	28·794
Ft. Stevenson, Dak...	1,784	e28·114	i28·142	28·110	e28·149	28·228	28·242	28·238	28·185	28·145	j28·141	c28·094	28·018	28·150
Ft. Macon, N. C.......	11	....	....	....	....	....	....	y30·125	30·151	29·813	29·950	30·034	29·922	......
Fredericksburg, Tex..	1,742	28·296	d28·304	28·277	j28·357	i28·393	j28·373	28·345	28·296	28·244	j28·231	28·206	28·224	28·295
Galveston, Tex........	40	29·984	29·941	29·968	30·040	30·095	30·095	30·096	30·026	29·958	29·966	29·924	29·981	30·002
Grand Haven, Mich...	620	29·320	29·368	29·368	29·365	29·444	29·308	29·424	29·415	29·193	29·338	29·304	29·271	29·353
Hatteras, N. C........	20	....	....	....	....	....	A30·029	30·112	30·141	29·786	29·983	30·029	29·928	......
Helena, Mont....... ..	4,202	25·861	25·827	25·807	25·893	25·945	25·782	25·809	25·743	25·769	25·745	25·775	25·717	25·811
Henrietta, Tex..	915	....	....	....	....	n29·198	29·136	29·115	i29·043	d28·942	k28·932	28·935	28·883	......
Indianapolis, Ind.....	753	29·222	29·239	29·276	29·283	29·329	29·302	29·305	29·275	29·031	29·178	29·211	29·126	29·231
Indianola, Tex........	26	29·996	29·948	29·973	30·070	30·132	30·128	30·135	30·030	29·957	29·964	29·981	29·935	30·017
Jacksborough, Tex...	1,133	....	f28·846	28·937	k28·975	29·018	28·973	28·957	28·881	28·806	28·792	e28·777	28·749	......
Jacksonville, Fla.....	43	30·038	29·989	30·031	30·044	30·135	30·075	30·113	30·101	29·920	30·000	29·968	29·947	30·080
Keokuk, Iowa.........	618	29·332	29·336	29·304	29·402	29·521	29·461	29·474	29·415	29·242	29·343	29·325	29·244	29·374
Key West, Fla........	27	30·058	29·972	29·994	29·990	30·076	j30·070	30·075	30·077	30·010	30·055	29·990	29·987	30·026
Kittyhawk, N. C......	22	30·012	30·048	30·062	30·112	30·211	30·037	30·132	30·164	29·798	29·944	30·006	29·933	30·043
Knoxville, Tenn.......	980	29·066	29·074	29·113	29·125	29·196	29·097	29·121	29·102	28·883	29·002	b29·045	28·990	29·068
La Crosse, Wis........	708	29·210	29·250	29·253	29·256	29·300	29·335	29·353	29·340	29·178	29·247	29·223	29·126	29·261
La Mesilla, N. M......	4,124	26·088	26·092	26·133	26·168	26·135	26·130	26·123	26·076	26·061	26·047	26·010	26·019	26·091
Laredo, Tex...........	401	29·493	29·490	29·535	29·690	29·704	29·675	e29·695	29·597	29·518	29·485	29·446	29·430	29·559
Leavenworth, Kan....	842	29·135	29·123	29·181	29·205	29·390	29·268	29·271	29·205	29·070	29·125	29·079	29·002	29·165
Lewiston, Idaho	800	29·169	29·122	29·232	29·324	29·517	29·179	29·361	29·227	29·240	29·173	29·177	29·079	29·238
Little Rock, Ark......	298	29·703	29·687	29·744	29·781	29·807	29·841	29·837	29·766	29·627	29·666	29·653	29·617	29·735
Los Angeles, Cal......	350	29·631	29·547	29·596	29·661	29·722	29·702	29·754	29·711	29·664	29·642	29·574	29·565	29·647
Louisville, Ky.........	530	29·457	29·464	29·510	29·523	29·635	29·542	29·553	29·530	29·282	29·411	29·439	29·367	29·476
Lynchburg, Va.......	652	29·341	29·400	29·413	29·419	29·541	29·363	29·416	29·471	29·116	29·262	29·374	29·256	29·370
Madison, Wis.........	949	28·981	29·012	29·013	29·011	29·083	29·036	29·062	29·047	28·872	28·986	28·907	28·912	29·001
Marquette, Mich......	673	29·253	29·314	29·230	29·237	29·308	29·256	29·344	29·207	29·210	29·277	29·309	29·228	29·281

Mason, Tex.	1,020	c28·363	b28·371	28·383	28·446	f28·492	28·464	28·438	28·387	b28·318	d28·324	28·399	28·302	28·391
Memphis, Tenn.	321	29·720	29·706	29·754	29·767	29·801	29·800	29·830	29·702	29·604	b29·673	29·672	29·692	29·739
Milwaukee, Wis.	697	29·283	29·278	29·270	29·294	29·308	29·298	29·348	29·358	29·137	29·261	29·279	29·198	29·273
Mobile, Ala.	69	30·081	j29·955	29·994	30·085	c30·188	30·070	30·078	30·064	29·910	29·976	29·989	29·912	30·019
Montgomery, Ala.	219	29·833	29·801	29·853	29·888	29·975	29·980	30·082	29·902	29·737	29·810	29·789	29·754	29·840
Moorhead, Minn	923	......	......	......	......	......	......	D29·140	29·090	29·040	29·016	28·969	28·881	......
Morgantown, W. Va.	963	29·030	29·085	29·109	29·105	29·187	29·052	29·100	29·123	28·798	28·951	29·043	28·987	29·048
Mt. Washington, N. H.	6,286	23·874	23·981	23·899	23·719	23·586	23·845	23·392	23·499	23·267	23·307	23·838	23·707	23·614
Nashville, Tenn.	507	29·538	29·516	29·571	29·585	29·690	29·607	29·614	29·677	29·370	29·478	29·491	29·481	29·587
New Haven, Conn	106	29·863	29·961	29·914	29·978	30·103	29·855	29·968	30·041	29·706	29·736	29·936	29·779	29·895
New London, Conn.	47	29·930	30·023	29·987	30·054	30·160	29·912	30·055	30·121	29·659	29·806	30·024	29·859	29·966
New Orleans, La.	52	30·005	c29·946	29·984	30·022	30·103	30·088	30·085	30·027	29·990	29·986	29·981	29·980	30·002
Newport, R. I.	84	29·938	30·021	29·977	30·043	30·162	29·901	30·026	30·000	29·699	29·771	30·007	29·844	29·950
New York city, N. Y.	164	29·891	29·904	29·880	29·989	30·060	29·837	29·940	30·002	29·507	29·700	29·880	29·787	29·857
Norfolk, Va.	30	29·951	30·009	30·019	30·068	30·181	30·008	30·100	30·187	29·782	29·884	30·000	29·891	29·998
North Platte, Neb.	2,811	27·100	27·087	27·125	27·137	27·170	27·114	27·110	27·007	27·027	27·005	27·067	26·978	27·068
New Shoreham, R. I.	37	......	......	m30·000	30·084	30·184	29·921	30·075	30·118	29·677	29·807	30·029	29·863	......
Olympia, Wash.	86	30·054	30·001	30·002	30·110	30·237	29·791	30·058	29·914	29·998	29·974	30·013	29·919	30·010
Omaha, Neb.	1,113	28·849	28·851	28·895	28·913	29·016	28·979	28·973	28·913	28·684	28·870	28·824	28·738	28·886
Oswego, N. Y.	304	29·653	29·733	29·697	29·744	29·830	29·664	29·778	29·818	29·437	29·568	29·692	29·583	29·683
Pembina, Dak	791	29·052	29·077	r......	......	......	......	......	......	......	......	......	......	......
Pensacola, Fla.	30	30·046	29·978	30·067	30·063	30·153	30·127	30·134	30·089	29·972	30·085	29·978	29·955	30·046
Philadelphia, Pa.	52	29·944	30·028	30·010	30·070	30·200	29·998	30·101	30·138	29·701	29·845	29·995	29·800	29·990
Phœnix, Ariz.	1,068	28·743	28·678	28·727	28·818	28·893	28·887	28·896	28·862	28·838	28·771	28·673	28·670	28·791
Pike's Peak, Col	14,134	18·081	18·010	17·955	17·811	17·496	17·539	17·439	17·506	17·548	17·733	17·840	17·974	17·739
Pilot Point, Tex.	800	f29·249	f29·221	29·203	f29·836	f29·403	c29·364	29·390	29·277	29·223	a......	......	......	......
Pioche, Nev.	6,280	24·092	24·069	24·114	24·133	24·072	23·997	24·041	24·031	24·005	24·000	23·981	23·996	24·046
Pittsburg, Pa.	762	29·200	29·245	29·265	29·281	29·371	29·238	29·288	29·318	28·978	29·130	29·219	29·126	29·221
Port Eads, La.	7	......	......	......	......	......	......	......	......	......	m30·054	29·998	29·990	......
Port Huron, Mich.	633	29·296	29·354	29·352	29·399	29·448	29·390	29·408	29·421	29·131	29·297	29·305	29·265	29·336
Portland, Me.	45	29·885	29·971	29·987	29·994	30·109	29·875	29·969	30·051	29·634	29·385	29·999	29·819	29·887
Portland, Or.	67	30·015	29·960	30·027	30·090	30·235	29·825	30·008	29·902	30·018	29·981	30·007	29·925	30·010
Prescott, Ariz.	5,300	24·810	24·786	24·814	24·828	24·774	24·759	24·758	24·757	24·728	24·747	24·709	24·754	24·769
Punta Rassa, Fla.	13	30·056	29·967	30·000	30·004	30·097	30·078	30·086	30·084	29·969	30·043	29·989	29·970	30·027
Red Bluff, Cal.	334	29·508	r29·452	29·560	29·673	29·898	29·643	29·707	29·758	29·701	29·617	29·552	29·525	29·659
Rio Grande City, Tex.	206	29·765	29·743	29·773	29·881	29·952	29·803	29·853	t29·851	29·787	29·746	29·717	29·707	29·814
Rochester, N. Y.	549	29·336	29·405	29·387	29·413	29·508	29·356	29·454	29·496	29·135	29·281	29·396	29·249	29·371
Roseburgh, Or.	537	29·521	29·471	29·516	29·603	29·732	29·839	29·539	29·507	29·543	29·498	29·501	29·450	29·519
Sacramento, Cal.	70	29·800	29·772	29·846	29·940	30·079	29·965	30·075	30·070	29·998	29·927	29·858	29·840	29·982
Salt Lake City, Utah.	4,348	25·694	25·594	25·685	25·741	25·759	25·694	25·081	25·657	25·046	25·605	25·570	25·548	25·644
San Antonio, Tex.	676	29·813	29·299	29·394	29·406	29·406	29·451	29·457	29·366	29·291	29·283	29·242	29·348	29·399
San Diego, Cal.	67	29·985	29·847	29·890	29·953	30·017	30·012	30·070	30·039	29·970	29·904	29·839	29·835	29·943

Monthly and Annual Mean Actual Barometer, July, 1880, to June, 1881, inclusive.—(Continued.)

STATION.	Elevation of cistern of barometer above m'n sea-level, feet.	1880. July.	1880. August.	1880. September.	1880. October.	1880. November	1880. December.	1881. January.	1881. February.	1881. March.	1881. April.	1881. May.	1881. June.	Annual mean.
Sandusky, O.	638	29·325	29·368	29·370	29·392	29·488	29·370	29·429	29·426	29·141	29·303	29·359	29·256	29·353
Sandy Hook, N. J.	28	29·945	30·027	30·005	30·086	30·206	29·982	30·093	30·143	29·708	29·853	30·022	29·881	29·995
San Francisco, Cal.	60	29·899	29·847	29·908	29·982	30·084	29·962	30·071	30·087	30·026	29·968	29·912	29·905	29·971
Santa Fé, N. M.	7,046	23·355	23·353	23·344	23·337	23·220	23·212	23·179	23·183	23·170	23·230	23·261	23·306	23·263
Savannah, Ga.	87	29·958	29·969	30·005	30·024	30·111	30·022	30·072	30·070	29·832	29·938	29·945	29·886	29·986
Shreveport, La.	227	29·794	29·785	29·807	29·804	29·951	29·916	29·907	29·928	29·745	29·753	29·731	29·729	29·817
Silver City, N. M.	5,796	24·321	24·311	24·268	24·106	24·025	24·116	24·242	24·238	24·208	24·233	24·211	24·273	24·213
Smithville, N. C.	34	30·023	30·060	30·074	30·098	30·192	30·005	30·133	30·154	29·833	29·975	30·031	29·930	30·047
Socorro, N. M.	4,565	25·432	25·442	25·480	25·443	25·452	25·432	25·393	25·382	25·356	25·350	J25·327	......	......
Springfield, Ill.	644	29·355	29·353	29·302	29·411	29·523	29·451	29·464	29·410	29·214	29·834	29·334	29·259	29·375
Springfield, Mass.	120	29·843	29·935	29·901	29·965	30·068	29·806	29·962	30·030	29·580	29·715	29·911	29·706	29·883
Spokane Falls, Wash.	1,910	......	......	......	......	......	......	......	K27·976	28·003	27·954	27·979	27·896	......
St. Louis, Mo.	568	29·440	29·439	29·493	29·516	29·636	29·560	29·570	29·505	29·312	29·421	29·417	29·345	29·471
St. Paul, Minn.	811	29·092	29·110	29·088	29·092	29·215	29·205	29·215	29·213	29·098	29·126	29·117	29·024	29·133
St. Vincent, Minn.	804	......	......	E29·008	29·074	29·205	29·247	29·271	29·240	29·107	29·133	29·088	29·009	......
Stockton, Tex.	3,050	26·963	26·965	27·009	27·041	27·048	27·022	27·010	i26·949	26·920	26·916	26·884	26·887	26·968
Thatcher's Isl'd, Mass.	48	29·922	30·010	29·971	30·067	30·154	29·898	30·016	30·083	29·637	29·758	30·012	29·841	29·947
Toledo, O.	651	29·306	29·351	29·362	29·373	29·465	29·357	29·407	29·404	29·131	29·287	29·341	29·238	29·335
Tucson, Ariz.	2,404	27·474	27·451	27·500	27·557	27·580	27·595	27·585	27·535	27·522	27·491	27·410	27·427	27·511
Umatilla, Or.	384	29·635	29·591	29·703	29·795	30·007	29·696	29·808	29·741	29·724	29·674	29·682	29·576	29·727
Uvalde, Tex.	891	29·055	j29·043	c29·080	j29·156	29·219	29·180	29·185	29·094	29·006	j28·984	e28·957	28·953	29·077
Vicksburg, Miss.	244	29·817	29·773	29·824	29·865	29·951	29·916	29·900	29·843	29·732	29·775	29·740	29·742	29·894
Virginia City, Mont.	5,810	24·329	24·278	24·331	24·355	L24·361	......	......	......	......	......	......	......	......
Visalia, Cal.	348	29·455	29·421	29·495	29·613	29·746	29·686	29·771	29·755	29·606	29·590	29·509	29·496	29·600
Washington, D. C.	105	29·894	29·960	29·967	30·020	30·143	29·961	30·057	30·082	29·685	29·829	29·941	29·814	29·946
Wilmington, N. C.	52	29·949	29·989	30·006	30·036	30·132	29·997	30·004	30·107	29·788	29·928	29·986	29·880	29·991
Winnemucca, Nev.	4,327	25·637	25·590	25·684	25·753	25·781	M25·750	25·674	25·647	25·636	25·598	25·577	25·519	25·653
Wood's Holl, Mass.	35	29·900	29·974	29·971	30·038	30·150	29·895	30·019	30·066	29·630	29·770	30·014	29·847	29·943
Yankton, Dak.	1,228	28·684	28·687	28·723	28·749	28·857	28·851	28·825	28·768	28·715	28·734	29·677	29·587	28·738
Yuma, Ariz.	141	29·649	29·602	29·674	29·787	29·920	29·892	29·954	29·867	29·819	29·746	29·648	29·624	29·765

Notes to Table on Pages 200–204.

- * Bell's "Climatology," etc., *loc. cit.*
- † Not determined.
- *a* Station closed November 30, 1880.
- *b* 20 days only.
- *c* 26 days only.
- *d* 23 days only.
- *e* 30 days only.
- *f* 24 days only.
- *g* 21 days only.
- *h* Observations discontinued Dec. 31, 1880.
- *i* 27 days only.
- *j* 28 days only.
- *k* 25 days only.
- *l* Station closed June 27, 1881.
- *m* Office and records burned Dec. 23, 1880; no observations taken Dec. 23, 24, and 25.
- *n* Observations commenced Oct. 13, 1880.
- *o* Barometer readings commenced Jan. 1, 1881.
- *p* Observations commenced Dec. 3, 1880.
- *q* No mail reports received prior to Aug. 1, 1880.
- *r* Record incomplete; observer sick.
- *s* Observations commenced Sept. 27, 1880.
- *t* Data incomplete owing to sickness of observer.
- *u* From 20th to 31st only.
- *v* Barometer unserviceable.
- *w* 17 days only.
- *x* 18 days only.
- *y* Observations commenced Jan. 3, 1881
- A Observations commenced Dec. 1, 1880.
- B Prior records incomplete.
- C 9 days only; office and records destroyed by fire Nov. 17, 1880.
- D Observations commenced Jan. 1, 1881.
- E Observations commenced Sept. 1, 1880.
- F Station closed Sept. 3, 1880.
- G Station closed Apr. 1, 1881.
- H Observations commenced Apr. 10, 1881.
- I August 16 to 31,
- J Station closed May 23, 1881.
- K Observations commenced Feb. 1, 1881.
- L Station removed to Eagle Rock, Idaho, on Nov. 19, 1880.
- M 3 days only; office and records destroyed by fire, Dec. 4, 1880.

Notes to Table on Pages 206–209.

- *a* Local observations discontinued Oct. 13, 1880.
- *b* For 15 days only.
- *c* Local observations discontinued Dec. 21, 1880.
- *d* For 9 days only.
- *e* Observations discontinued Nov. 30, 1880.
- *f* Observations discontinued Dec. 31, 1880.
- *g* Observations commenced Oct. 13, 1880.
- *h* For 30 days only.
- *i* Office burned Dec. 23, 1880; no observations taken Dec. 23, 24, and 25.
- *j* Observations commenced Dec. 3, 1880.
- *k* Local observations commenced Jan. 1, 1881.
- *l* Local observations discontinued March 31, 1881.
- *m* Observations commenced Jan. 3, 1881.
- *n* Observations commenced Dec. 1, 1880.
- *o* For 28 days only.
- *p* For 29 days only.
- *q* Local observations discontinued July 31, 1880.
- *r* For 9 days only; office destroyed by fire on Nov. 17, 1880.
- *s* Observations commenced Jan. 1, 1881.
- *t* For 30 days only.
- *u* Observations commenced Sept. 1, 1880.
- *v* For 20 days only.
- *w* Station closed Sept. 3, 1880.
- *x* For 25 days only.
- *y* For 26 days only.
- *z* For 24 days only.
- A For 18 days only.
- B For 21 days only.
- C For 23 days only.
- D August 16th to 31st only.
- E Observations commenced April 1, 1881.
- F Local observations commenced Jan. 1, 1881.
- G Local observations discontinued March 19, 1881.
- H Station removed to Eagle Rock, Idaho, on Nov. 19, 1880.
- I For 3 days only; office and records destroyed by fire; observations recommenced Dec. 28, 1880.
- J Local observations commenced Feb. 1, 1881.
- K Observations commenced Sept. 4, 1880.

Monthly and Annual Mean Temperature, from Observations taken at 7 A. M., 2 and 9 P. M. (Local Time), from July, 1880, to June, 1881, inclusive.

The daily means are obtained by dividing the sum of the 7 A. M., 2 P. M., and twice the 9 P. M. observations by four; the monthly means by dividing the sum of the daily means by the number of days in the month.—(U. S. Signal Service.)

STATION.	Elevation of thermometer above-ground.	1880. July.	August.	September.	October.	November	December.	1881. January.	February.	March.	April.	May.	June.	Annual mean.
	Feet.													
Albany, N. Y.	51	75·3	71·2	65·5	51·8	38·1	25·1	20·2	27·2	30·0	46·9	65·6	65·4	49·1
Alpena, Mich	54	65·7	65·2	57·6	45·0	26·8	19·7	14·0	15·9	26·3	24·3	54·0	56·3	40·1
Atlanta, Ga.	78	79·7	76·5	69·4	61·0	48·1	42·6	40·8	46·8	40·3	59·5	71·1	78·6	60·8
Atlantic City, N. J.	10	72·1	71·7	67·9	56·1	41·8	29·2	27·7	30·3	38·4	44·7	57·1	64·7	50·2
Augusta, Ga.	18	82·2	80·1	74·0	64·0	51·1	45·2	42·5	50·8	53·0	62·2	75·1	81·7	63·5
Baltimore, Md.	33	77·6	74·8	68·4	56·0	42·1	31·6	30·8	34·8	42·0	52·0	67·9	71·2	54·1
Barnegat, N. J.	6	74·4	70·6	66·1	54·7	40·2	27·9	26·7	29·4	37·5	43·3	55·9	63·2	49·2
Bismarck, Dak.	16	69·5	66·6	55·4	42·3	19·7	3·6	—1·3	9·6	23·7	37·9	60·8	66·3	37·8
Boisé City, Idaho	19	75·0	71·9	61·3	51·3	32·2	35·4	31·6	30·2	45·4	54·8	60·9	67·1	52·2
Boston, Mass	156	71·1	68·9	64·1	50·8	37·5	26·2	22·4	27·9	36·6	43·4	55·2	60·6	47·1
Brackettville, Tex	6	81·8	77·5	75·6	*a*68·4	*b*42·8	*c*55·6	....	*d*58·6	*b*63·9	....	....	....	....
Breckinridge, Minn.	6	69·7	64·8	54·1	42·2	*e*17·8	....	....	....	....	....	....	....	....
Buffalo, N. Y.	72	70·8	69·2	62·9	49·1	32·6	22·4	18·8	21·8	29·5	36·8	57·0	59·8	44·2
Burlington, Vt.	55	71·7	68·4	62·9	48·7	33·8	20·4	14·1	22·2	33·0	41·2	50·6	61·6	44·8
Cairo, Ill.	44	79·5	79·8	67·5	58·3	37·7	33·0	29·9	38·3	46·0	56·7	72·6	77·9	56·4
Cape Hatteras, N. C.	7	79·1	78·9	70·1	62·7	*e*55·9	....	....	....	....	....	....	....	....
Cape Henry, Va.	16	78·4	76·4	72·1	62·7	49·2	37·6	35·4	38·5	44·7	52·0	65·5	72·9	57·1
Cape Lookout, N. C.	18	80·9	79·1	74·4	65·1	53·5	*f*43·0	....	....	....	....	....	....	....
Cedar Keys, Fla.	20	82·6	80·4	78·7	70·6	63·1	54·9	54·8	58·3	58·8	66·2	76·7	82·4	69·0
Champaign, Ill	....	....	....	....	*g*45·9	29·5	22·1	19·5	26·4	33·6	45·9	68·3	69·9	....
Charleston, S. C.	40	83·4	81·4	75·8	65·9	53·5	*h*48·6	47·0	52·3	54·6	60·3	73·1	81·8	64·8
Chattanooga, Tenn.	43	77·0	77·0	67·6	59·6	45·3	37·7	37·6	44·4	47·2	57·9	71·4	77·0	58·3
Charlotte, N. C	85	78·9	76·3	69·7	59·2	45·5	38·4	37·2	48·6	47·5	55·8	70·9	78·7	58·5
Cheyenne, Wyo.	15	66·8	64·8	56·9	42·8	20·2	27·9	23·9	28·8	33·4	40·3	54·3	67·7	44·5
Chicago, Ill.	69	72·7	72·9	63·1	51·3	31·6	23·2	20·0	25·1	32·4	41·7	60·7	73·2	46·5
Chincoteague, Va.	29	74·7	72·7	68·7	59·4	44·1	32·0	30·7	32·8	40·3	45·7	57·9	67·1	52·2

Cincinnati, Ohio	68	77·1	76·8	67·6	56·3	38·3	31·2	30·6	36·5	42·1	51·2	71·6	74·0	54·4
Cleveland, Ohio	73	71·1	70·9	63·7	50·7	31·6	28·8	20·5	26·3	31·8	40·2	62·3	64·3	46·3
Columbus, Ohio	52	73·4	74·7	66·5	52·0	33·7	26·1	25·0	29·5	37·4	47·6	68·2	70·5	50·4
Corsicana, Tex.	19	82·8	82·4	74·8	63·6	45·9	f47·4	40·2	49·5	58·4	68·4	76·2	86·2	64·6
Davenport, Ia.	46	75·4	75·1	63·5	51·4	31·0	22·2	14·2	20·1	29·2	46·3	69·6	70·8	47·8
Deadwood, Dak.	16	66·7	64·7	55·4	42·5	20·1	17·8	15·9	24·1	32·0	41·0	54·2	63·5	40·5
Del. Breakwater, Del.	13	73·6	72·4	68·7	58·8	44·0	31·8	29·5	32·5	30·1	45·8	59·2	66·2	51·8
Denison, Tex.	17	79·7	80·0	73·1	61·6	41·3	41·1	33·7	43·6	53·2	64·6	78·7	84·4	60·8
Denver, Col.	73	71·3	69·7	61·7	47·3	21·7	30·2	26·0	29·6	37·8	58·5	60·1	73·5	48·6
Des Moines, Ia.	35	74·4	73·1	62·2	49·2	28·2	h20·1	14·4	20·7	29·1	44·2	67·4	71·3	46·2
Detroit, Mich.	61	70·8	69·3	61·6	48·4	30·1	28·2	17·7	26·5	34·2	44·9	64·7	65·3	46·3
Dodge City, Kan.	15	76·4	74·8	65·8	51·6	26·2	24·9	20·1	26·6	30·9	54·0	64·4	78·5	50·3
Dubuque, Ia.	32	74·2	72·8	61·2	48·7	27·8	19·1	13·1	20·3	31·2	45·0	68·7	68·9	45·9
Duluth, Minn.	19	65·0	64·0	54·9	43·3	19·9	12·9	5·5	15·7	28·2	38·1	51·1	56·8	38·0
Eagle Rock, Idaho	12	.....	.....	.....	.....	.....	j28·8	18·0	26·2	36·5	48·4	55·9	63·6	.....
Eastport, Me.	32	61·8	60·7	57·2	48·0	33·9	24·4	18·1	25·2	33·4	38·1	47·1	52·4	41·7
El Paso, Tex.	17	.....	.....	.....	.....	.....	.....	k40·3	51·3	l55·3	.....	.....	.....	.....
Erie, Pa.	38	72·7	70·8	63·0	51·0	34·4	25·3	28·0	25·4	32·3	40·5	60·5	61·2	46·7
Escanaba, Mich.	25	65·8	65·2	57·4	44·0	23·5	17·2	7·6	13·2	24·0	32·6	54·5	59·0	38·7
Ft. Apache, Ariz.	7	.....	.....	.....	.....	.....	.....	k33·0	40·6	43·0	57·3	63·5	74·0	.....
Ft. Gibson, Ind. Ter.	19	79·0	79·1	71·0	58·0	36·5	35·5	27·5	36·7	47·9	61·2	71·6	81·9	57·2
Ft. Macon, N. C.	8	.....	.....	.....	.....	.....	.....	m41·9	44·7	49·4	54·0	68·5	75·4	.....
Galveston, Tex.	36	83·1	82·6	78·0	69·6	54·1	58·5	47·2	54·8	61·2	68·1	77·5	85·0	67·9
Grand Haven, Mich.	28	69·9	68·7	60·9	49·2	32·3	26·3	19·7	23·3	30·9	40·6	62·4	68·0	46·5
Hatteras, N. C.	6	.....	.....	.....	.....	.....	n41·9	40·0	41·4	47·7	52·6	67·0	74·0	.....
Indianapolis, Ind.	52	76·6	76·2	65·1	53·3	31·7	24·3	24·2	30·3	37·3	46·0	71·3	73·4	51·0
Indianola, Tex.	29	82·7	81·9	79·9	69·3	52·8	53·9	45·4	55·3	63·8	69·8	77·7	85·0	68·1
Jacksonville, Fla.	37	82·6	81·2	76·9	69·0	61·5	54·9	53·0	58·0	60·0	67·4	75·8	83·2	68·6
Keokuk, Ia.	47	78·8	78·5	66·3	52·5	31·6	24·6	19·1	25·2	34·0	47·5	70·8	73·0	50·1
Key West, Fla.	48	85·5	83·3	83·1	79·2	78·3	o69·9	72·3	70·9	70·2	74·6	80·2	84·7	77·7
Kittyhawk, N. C.	4	79·1	76·4	72·4	62·4	51·8	38·5	36·8	39·5	44·8	50·9	63·7	72·7	57·4
Knoxville, Tenn.	72	75·2	75·9	67·0	58·5	42·6	34·4	35·5	41·8	44·0	54·6	p70·6	74·0	56·2
La Crosse, Wis.	40	73·0	71·1	60·7	48·5	25·5	15·8	9·6	18·9	30·8	43·5	68·2	69·3	44·6
La Mesilla, N. M.	17	q78·7	.....	.....	.....	.....	.....	.....	.....	.....	.....	.....	.....	.....
Leavenworth, Kan.	34	78·2	77·6	65·6	52·7	31·9	25·8	21·1	25·5	36·8	50·8	70·1	77·6	51·1
Little Rock, Ark.	26	78·8	77·4	69·0	60·9	41·8	40·8	36·1	44·8	51·9	62·7	72·7	81·7	59·9
Los Angeles, Cal.	37	63·5	66·1	63·7	61·6	54·8	55·0	51·7	57·5	56·1	61·2	62·6	65·3	59·9
Louisville, Ky.	89	77·8	77·9	66·9	56·2	37·5	30·5	29·8	36·5	41·7	52·4	72·2	74·8	54·5
Lynchburg, Va.	30	78·4	76·3	67·9	57·1	42·1	34·1	34·0	38·8	43·9	54·4	70·6	76·0	56·1
Madison, Wis.	33	71·9	71·0	60·6	48·5	26·8	17·0	12·6	18·5	28·5	40·9	66·2	66·4	44·0
Marquette, Mich.	36	64·4	62·3	55·9	44·0	24·7	19·3	10·1	14·3	24·8	35·2	53·3	55·1	38·6
Memphis, Tenn.	53	79·2	78·7	68·6	59·9	39·9	37·3	34·0	43·1	50·5	j60·0	74·1	81·5	58·9

Monthly and Annual Mean Temperature, July, 1880, to June, 1881, inclusive.—(Continued.)

STATION.	Elevation of thermometer above-ground.	1880. July.	1880. August.	1880. September.	1880. October.	1880. November	1880. December.	1881. January.	1881. February.	1881. March.	1881. April.	1881. May.	1881. June.	Annual mean.
	Feet.													
Milwaukee, Wis.	105	68·9	69·6	61·1	48·4	27·7	19·6	16·6	23·0	30·1	40·8	58·3	60·4	43·7
Mobile, Ala.	64	80·0	*o*80·3	75·2	66·7	*r*52·3	40·2	47·9	53·7	57·0	65·6	76·9	83·0	65·6
Montgomery, Ala.	34	81·9	80·5	73·6	65·6	51·7	46·9	45·2	51·1	54·3	64·4	76·2	82·9	64·5
Moorhead, Minn.	23	......	......	......	......	......	......	—2·6	6·8	10·3	34·2	60·7	65·8	.....
Morgantown, W. Va.	10	71·8	71·4	63·5	53·7	40·1	31·1	32·2	34·9	41·7	40·4	66·0	68·5	52·0
Mt. Washington, N.H.	28	47·6	47·0	41·5	30·0	14·2	7·8	2·4	7·2	14·7	16·1	39·0	38·8	25·5
Nashville, Tenn.	34	78·4	79·4	68·2	59·9	41·2	35·0	34·9	41·0	46·0	57·8	73·9	78·7	58·1
New Haven, Conn.	112	73·5	70·1	64·3	52·0	40·0	27·8	21·7	26·9	36·1	44·1	58·5	62·5	48·1
New London, Conn.	29	71·7	69·5	64·2	52·2	39·5	28·2	24·7	29·1	37·5	44·1	56·7	61·8	48·3
New Orleans, La.	45	81·7	*t* 81·3	76·8	68·0	56·4	53·0	50·4	56·3	59·0	67·2	77·0	84·8	67·7
Newport, R. I.	19	71·5	69·2	64·3	53·5	41·1	20·7	26·5	29·7	37·5	42·7	54·0	61·2	48·4
New York city, N. Y.	148	73·5	71·9	65·7	54·0	40·1	27·7	26·0	29·8	37·1	46·1	60·4	64·2	49·7
New Shoreham, R. I.	8	......	......	*u*64·8	54·5	43·0	31·9	28·8	31·6	37·0	42·5	53·0	60·3	......
Norfolk, Va.	20	80·2	76·8	71·3	60·6	47·0	36·2	34·5	30·6	45·6	52·5	67·1	74·4	57·1
North Platte, Neb.	19	73·2	72·5	61·2	47·8	23·7	18·0	13·5	21·7	34·3	48·3	62·2	71·9	45·7
Olympia, Wash.	23	62·3	60·8	54·5	49·3	*v*39·2	39·8	38·8	43·8	46·6	50·9	53·0	58·0	49·7
Omaha, Neb.	59	77·7	75·7	63·6	49·4	26·7	18·4	11·8	18·6	28·1	45·1	68·4	75·9	46·2
Oswego, N. Y.	85	71·1	69·0	63·6	50·7	36·8	25·6	21·1	25·2	33·7	40·8	58·4	58·5	46·6
Pembina, Dak.	6	65·5	*w*62·3	......	......	......	......	......	......	......	......	......	......	.....
Pensacola, Fla.	20	81·0	80·7	76·0	67·5	56·5	51·5	40·5	55·2	57·9	64·8	76·3	82·4	66·6
Philadelphia, Pa.	99	75·6	72·8	67·7	55·1	40·5	28·6	25·9	30·7	38·8	48·7	64·1	67·4	51·3
Pike's Peak, Col.	5	38·7	37·4	30·9	18·3	0·5	6·7	Zero.	5·3	5·0	18·7	25·6	40·2	18·9
Pioche, Nev.	5	73·5	72·3	65·1	51·0	32·0	35·1	20·8	37·3	39·4	54·6	60·6	70·2	51·7
Pittsburg, Pa.	88	73·4	73·7	65·8	53·4	35·9	26·8	28·2	30·1	37·1	47·4	66·7	68·7	50·6
Port Huron, Mich.	80	68·8	67·8	60·3	47·9	28·9	21·8	16·4	21·9	29·0	37·2	56·8	58·0	42·9
Portland, Me.	28	70·5	68·1	62·4	50·5	37·8	27·8	22·4	28·6	38·1	43·9	54·8	60·5	47·1
Portland, Ore.	45	66·1	63·8	59·8	52·7	42·2	39·7	39·0	45·5	40·2	55·1	58·1	60·8	52·7
Portsmouth, N. C.	8	*x*70·3	*y*77·6	*z*73·6	A66·3	B50·4	C41·4	*x*39·1	42·8	49·6	53·4	68·2	75·4	59·8
Prescott, Ariz.	10	72·6	71·4	64·4	52·3	36·3	37·8	34·7	40·8	40·9	56·8	62·2	71·3	53·5
Punta Rassa, Fla.	14	81·8	80·4	80·3	75·1	73·2	63·2	64·8	65·5	65·4	70·0	76·5	82·1	73·2
Red Bluff, Cal.	21	85·4	D78·8	75·2	65·5	49·8	49·1	49·4	52·9	55·8	64·2	70·5	74·6	64·3

Rochester, N. Y.	100	71·3	69·3	63·2	49·4	33·7	26·6	20·7	24·0	32·0	41·8	68·1	61·4	46·0
Roseburg, Ore.	20	68·8	65·5	61·7	51·4	39·2	46·6	42·8	47·4	49·0	55·7	57·9	60·2	58·8
Sacramento, Cal.	38	72·9	71·3	68·9	62·7	49·6	50·0	49·2	58·6	56·0	62·0	66·7	66·0	60·9
Salt Lake City, Utah.	52	75·3	73·5	64·1	51·5	30·3	34·3	38·1	38·2	41·9	54·5	61·5	71·6	52·4
San Antonio, Tex.	17	83·0	79·0	76·6	65·4	45·8	50·7	43·8	54·0	63·4	69·4	76·4	86·2	66·2
San Diego, Cal.	10	63·2	65·6	62·8	60·9	55·6	55·4	52·4	54·9	54·2	60·7	62·4	64·2	59·4
Sandusky, Ohio.	54	72·9	72·5	64·2	51·3	32·9	24·8	26·0	27·0	33·3	43·2	64·3	66·8	48·0
Sandy Hook, N. J.	15	74·4	71·6	67·6	56·2	43·1	29·6	26·3	29·2	37·4	45·7	60·5	65·2	50·6
San Francisco, Cal.	48	57·7	58·2	58·1	58·0	54·1	53·2	53·8	55·0	54·0	57·1	56·8	57·7	56·2
Santa Fé, N. M.	21	67·3	64·9	57·7	46·3	29·4	29·9	24·0	33·0	37·6	52·6	60·0	71·4	47·9
Savannah, Ga.	41	83·1	81·2	75·1	65·7	56·1	50·5	48·5	53·8	56·7	63·3	74·2	82·6	65·9
Shreveport, La	33	80·6	81·2	73·1	63·5	46·2	46·6	41·2	48·9	56·1	67·6	75·6	85·3	66·8
Smithville, N. C.	18	82·2	80·4	74·5	64·9	53·0	44·8	43·6	47·4	51·5	56·8	70·2	78·5	62·3
Springfield, Ill	39	77·1	77·2	64·6	53·6	32·7	30·2	28·4	30·5	36·2	48·7	70·3	72·2	50·9
St. Michael's, Alaska.	......	52·6	52·3	41·2	28·3	24·4	11·3	17·6	1·2	3·3	23·8	31·9	46·6	27·9
Springfield, Mass	54	73·3	69·8	64·4	50·5	37·6	26·3	20·9	26·8	36·8	46·1	61·8	64·3	46·2
St. Louis, Mo	104	77·8	78·4	66·6	54·9	32·1	27·3	28·0	30·3	38·9	51·4	71·5	75·5	52·2
St. Paul, Minn	82	71·6	69·5	58·9	45·7	22·6	13·6	7·9	17·2	30·7	44·0	64·5	67·9	42·8
Sitka, Alaska.	64	......	......	......	......	......	......	......	......	......	E42·5	45·4	51·2	......
St. Vincent, Minn.	9	......	......	K51·8	40·1	15·7	1·9	—5·6	5·8	19·6	31·8	58·7	62·6	......
Socorro, N. M.	5	......	......	......	......	......	......	F28·6	42·7	G43·2	......	......	......	......
Thatcher's Isl'd, Mass.	7	69·9	65·6	61·4	50·2	38·6	27·9	24·3	28·4	37·5	41·9	52·7	58·9	46·2
Toledo, Ohio.	64	73·8	72·2	64·4	51·7	34·0	25·5	22·8	26·9	35·0	46·2	66·3	67·1	48·7
Umatilla, Ore.	7	73·5	70·2	63·9	53·3	35·5	29·6	25·3	34·6	48·9	56·3	60·5	66·1	51·5
Vicksburg, Miss	32	80·7	80·1	71·9	63·7	47·7	47·0	43·2	50·3	56·2	60·9	76·8	84·6	64·1
Virginia City, Mont.	24	62·7	61·3	53·5	41·4	H22·3	......	......	......	......	......	......	......	......
Visalia, Cal.	22	80·8	77·8	73·8	61·2	46·7	49·9	48·1	53·9	55·7	63·2	70·6	74·1	63·0
Washington, D. C.	44	77·2	75·1	68·1	55·5	40·7	30·6	28·2	33·2	40·7	50·8	67·9	70·9	53·2
Wilmington, N. C.	28	79·4	77·4	73·0	63·4	51·7	48·3	43·1	40·1	52·1	57·8	70·7	78·4	61·6
Winnemucca, Nev.	7	75·1	70·3	62·8	48·9	30·3	I30·9	34·9	39·9	41·7	54·2	59·0	66·8	52·0
Wood's Holl, Mass.	6	70·2	68·4	64·3	51·4	42·3	30·9	27·2	30·1	37·5	42·2	53·7	59·9	48·4
Yankton, Dak	20	70·2	74·2	62·8	47·4	23·9	14·6	6·9	14·6	28·4	39·5	60·8	72·7	43·6
Yuma, Ariz.	5	......	......	......	......	......	......	......	J61·4	63·5	73·5	79·4	86·6	......

Monthly and Annual Mean Relative Humidity, from Observations taken at 7 A. M., 2 and 9 P. M. (Local Time) —July, 1880, to June, 1881, inclusive.

The daily means are obtained by dividing the sum of the 7 A. M., 2 and 9 P. M. observations by three; the monthly means by dividing the sum of the daily means by the number of days in the month.—(U. S. Signal Service.)

STATION.	1880.						1881.						Annual mean.
	July.	August.	September.	October.	November	December.	January.	February.	March.	April.	May.	June.	
Albany, N. Y.	59·4	63·1	67·8	65·4	67·8	74·8	70·1	69·7	69·0	54·1	62·9	62·8	65·6
Alpena, Mich.	75·3	71·3	74·7	78·5	80·2	80·7	75·2	77·4	81·0	68·9	74·4	69·3	75·2
Atlanta, Ga.	62·9	73·8	73·6	66·9	65·1	67·7	71·7	58·5	53·8	56·8	61·8	60·3	64·4
Atlantic City, N. J.	82·0	83·9	78·5	75·6	74·8	76·6	76·8	83·8	74·6	73·6	85·2	79·8	78·7
Augusta, Ga.	65·3	71·4	68·6	74·4	80·3	74·2	76·8	65·5	60·2	60·7	58·8	60·7	68·1
Baltimore, Md.	64·1	70·6	66·1	66·8	66·1	69·2	71·6	66·7	69·2	58·9	64·3	68·1	66·3
Barnegat, N. J.	79·8	81·2	78·0	74·9	73·8	78·6	78·5	78·0	75·1	73·3	86·5	80·8	78·2
Bismarck, Dak.	59·9	63·2	65·8	63·6	78·7	89·4	92·4	86·8	82·2	70·8	56·7	65·4	72·9
Boisé City, Idaho	83·7	38·2	88·6	44·9	59·3	75·5	73·0	73·6	65·2	55·8	46·3	43·6	54·0
Boston, Mass.	70·6	70·9	69·7	66·2	63·9	68·9	65·7	71·4	77·0	56·7	74·2	69·3	68·7
Brackettville, Tex.	67·8	74·6	74·9	*a* 67·2	*b* 74·3	*c* 74·3	....	*d* 54·3	*b* 51·5	....	....	....	......
Breckinridge, Minn.	69·4	73·4	70·0	70·4	77·8	(*e*)	....	....	....	....	....	....	......
Buffalo, N. Y.	67·6	69·3	68·8	69·8	75·1	80·9	79·9	76·4	80·8	70·7	71·3	78·1	78·6
Burlington, Vt.	61·9	62·7	68·6	67·5	70·4	77·1	70·2	73·3	70·8	55·8	70·6	63·7	67·7
Cairo, Ill.	69·3	70·6	73·5	71·8	71·8	75·4	76·2	72·2	62·5	63·3	71·0	71·7	70·8
Cape Hatteras, N. C.	80·3	81·7	77·0	74·8	*e* 79·2	....	....	....	....	....	....	....	......
Cape Henry, Va.	72·8	76·2	71·0	65·4	71·5	69·4	74·3	73·7	66·6	66·9	75·0	72·1	71·2
Cape Lookout, N. C	74·8	78·3	74·3	74·1	80·8	*f* 78·2	....	....	....	....	....	....	......
Cedar Keys, Fla.	72·7	76·9	75·0	75·0	83·4	78·8	81·0	75·2	66·8	69·8	69·0	69·8	74·4
Champaign, Ill.	....	....	....	*g* 65·3	64·0	70·3	66·2	70·3	72·1	67·7	65·3	76·7	......
Charleston, S. C.	70·7	72·8	71·5	80·2	77·3	*h* 71·7	77·6	66·2	62·6	66·9	71·9	67·5	71·4
Charlotte, N. C.	66·7	74·4	68·7	62·5	70·5	70·6	72·4	63·4	58·5	60·0	59·2	55·2	65·2
Chattanooga, Tenn.	71·7	75·4	76·9	73·5	67·3	72·0	73·9	64·2	59·6	60·6	66·7	70·1	69·3
Cheyenne, Wyo.	43·9	47·0	38·6	45·0	56·1	49·1	57·5	50·4	52·9	51·2	52·6	83·0	48·1
Chicago, Ill.	72·2	70·6	66·9	64·9	65·7	71·9	68·6	72·1	75·3	65·1	64·8	73·2	69·3
Chincoteague, Va.	81·2	84·7	79·5	76·4	70·1	80·9	84·1	79·7	75·4	77·9	89·9	81·5	81·1
Cincinnati, Ohio	63·5	66·7	69·4	68·4	67·9	70·1	69·2	67·8	65·3	62·8	58·8	65·9	65·4

Cleveland, Ohio	70·0	71·8	67·2	67·5	75·9	86·1	82·3	77·8	83·5	74·3	69·0	71·8	74·5
Columbus, Ohio	59·9	63·2	64·0	65·6	68·3	73·9	75·3	72·3	70·6	62·7	60·1	62·4	66·5
Corsicana, Tex	63·9	61·0	60·9	71·9	63·8	*f* 62·0	68·9	62·1	51·3	53·9	68·4	57·5	62·5
Davenport, Ia	68·0	67·3	67·0	62·0	62·9	70·9	72·5	77·6	75·1	68·3	57·3	69·1	68·2
Deadwood, Dak	53·8	57·4	50·6	52·7	60·8	62·9	65·5	66·1	63·9	61·6	60·1	56·8	59·3
Del. Breakwater, Del	79·6	83·4	75·8	69·2	75·8	78·3	80·9	78·3	77·2	75·0	84·1	79·7	78·1
Denison, Tex	73·2	68·6	72·1	65·7	71·8	75·1	76·7	73·2	61·3	59·0	75·6	64·4	69·6
Denver, Col	47·0	47·2	43·9	55·9	64·3	62·3	58·5	56·0	56·2	45·2	52·2	31·3	51·7
Des Moines, Ia	69·1	70·3	69·5	65·4	66·6	*h* 72·1	71·2	77·5	74·6	72·3	69·0	75·7	71·1
Detroit, Mich	70·8	77·3	72·6	73·8	71·4	76·1	77·2	78·4	74·7	61·7	62·3	65·6	71·8
Dodge City, Kan	60·7	58·9	53·7	56·5	60·7	63·6	58·8	65·7	55·6	52·1	75·8	51·8	50·7
Dubuque, Ia	66·2	69·8	70·7	65·3	60·1	74·7	70·3	74·2	70·3	63·2	59·2	70·7	60·1
Duluth, Minn	69·0	73·8	67·1	74·0	85·7	87·8	79·8	79·4	73·9	60·6	74·0	73·9	75·7
Eagle Rock, Idaho	......	......	......	......	......	*j* 77·6	77·7	78·5	63·1	50·0	39·2	37·4	......
Eastport, Me	70·0	78·1	70·7	75·5	67·8	71·7	67·9	70·5	70·4	61·8	76·9	74·2	72·8
El Paso, Tex	......	......	......	......	......	......	*k* 43·6	32·8	*l* 22·7	......	......	......	......
Erie, Pa	62·7	67·6	68·3	67·8	72·4	80·7	83·4	78·5	81·3	71·0	71·5	70·8	73·5
Escanaba, Mich	73·6	72·8	69·3	70·4	77·8	82·7	79·3	78·9	78·3	68·8	71·4	70·6	75·0
Ft. Apache, Ariz	......	......	......	......	......	......	*k* 41·5	43·7	49·4	40·8	27·2	10·1	......
Ft. Gibson, Ind. Ter	66·1	65·3	61·6	62·0	68·0	65·6	70·0	70·2	56·7	55·6	77·8	68·1	65·7
Ft. Macon, N. C	......	......	......	......	......	......	*m* 88·1	78·5	72·0	77·0	79·8	82·8	......
Galveston, Tex	72·2	72·2	79·4	70·3	81·5	80·1	79·7	79·2	69·9	75·0	75·2	68·8	75·3
Grand Haven, Mich	76·8	78·1	73·8	72·9	76·8	82·5	81·1	88·1	81·2	72·4	65·1	71·6	76·8
Hatteras, N. C	......	......	......	......	......	*n* 70·8	84·5	81·4	74·9	70·4	80·3	82·2	......
Indianapolis, Ind	67·0	67·6	62·2	65·0	67·6	71·8	73·6	78·8	71·1	64·0	55·4	66·3	67·1
Indianola, Tex	75·8	76·8	77·2	71·2	81·3	80·5	79·5	78·6	70·8	75·1	76·6	68·9	76·0
Jacksonville, Fla	67·5	71·9	73·0	73·2	77·7	64·0	73·4	62·7	54·7	59·0	68·5	64·2	67·5
Keokuk, Ia	62·3	61·6	64·8	62·9	66·1	72·6	72·0	74·4	71·7	68·8	62·3	70·5	67·5
Key West, Fla	67·5	73·0	73·6	73·6	76·9	*o* 75·9	80·7	72·5	69·0	68·3	70·7	71·6	72·8
Kittyhawk, N. C	75·7	80·2	76·3	73·8	77·1	74·7	81·3	81·1	68·2	72·7	81·1	77·2	76·6
Knoxville, Tenn	73·9	74·1	73·6	70·3	71·9	74·7	79·5	65·8	64·6	63·0	*p* 61·1	71·9	70·4
La Crosse, Wis	68·6	69·1	67·7	65·3	73·1	75·5	75·8	72·4	69·8	61·3	58·9	66·7	68·6
La Mesilla, N. M	*q* 48·2	......	......	......	......	......	......	......	......	......	......	......	......
Leavenworth, Kan	61·6	62·2	67·0	62·2	67·0	69·7	70·6	74·8	65·2	62·3	68·8	66·5	66·5
Little Rock, Ark	75·1	78·6	80·1	72·6	75·7	74·4	81·8	78·5	59·0	65·0	79·1	71·8	73·9
Los Angeles, Cal	73·4	71·9	71·3	62·5	55·1	71·6	62·9	62·1	64·2	71·2	69·1	68·4	67·0
Louisville, Ky	63·3	69·0	70·4	69·5	66·6	69·9	76·4	71·4	66·3	64·2	62·8	68·0	68·1
Lynchburg, Va	64·5	70·0	67·0	61·7	67·5	67·4	71·7	66·4	55·9	54·9	57·9	59·9	63·7
Madison, Wis	71·6	74·1	70·9	66·8	72·0	74·7	73·9	76·3	74·5	71·4	67·8	74·8	72·4
Marquette, Mich	66·9	70·8	65·5	70·0	68·6	72·8	68·4	66·2	73·5	58·0	66·6	68·6	68·0
Memphis, Tenn	71·8	72·6	77·3	78·1	79·4	79·6	77·7	71·5	57·4	*p* 64·1	66·4	66·6	71·9
Milwaukee, Wis	75·6	74·7	68·1	67·8	73·4	78·0	78·1	82·2	79·1	68·3	69·3	72·6	73·0

Monthly and Annual Mean Relative Humidity, July, 1880, to June, 1881, inclusive.—(Continued.)

STATION.	1880.						1881.						Annual mean.
	July.	August.	September.	October.	November	December.	January.	February.	March.	April.	May.	June.	
Mobile, Ala.	76·7	*o* 79·8	79·5	77·7	*r* 85·7	73·6	77·7	69·4	59·5	68·1	66·8	71·4	73·8
Montgomery, Ala.	65·1	69·7	70·6	68·9	69·6	70·6	75·7	62·0	57·2	59·6	58·8	60·1	65·7
Moorhead, Minn.	.....	.....	.....	.....	.....	.....	*s* 68·7	74·9	73·7	72·0	67·2	71·4	.....
Morgantown, W. Va.	75·8	78·5	77·0	69·8	63·6	65·3	73·4	66·3	63·6	65·5	69·2	74·2	70·2
Mt. Washington, N. H.	81·9	80·3	86·6	80·7	88·6	78·9	74·2	78·9	86·3	79·6	82·1	78·3	81·4
Nashville, Tenn.	67·8	66·4	72·0	70·7	71·7	73·9	78·0	66·1	61·2	63·7	63·6	63·3	68·2
New Haven, Conn.	73·5	76·8	78·8	73·8	64·9	70·9	66·9	70·2	68·2	56·9	74·9	69·6	70·4
New London, Conn.	80·9	79·1	78·2	74·5	69·9	72·5	69·8	72·5	74·5	64·4	80·6	78·3	74·6
New Orleans, La.	73·0	*t* 73·5	76·2	73·3	76·0	70·5	73·3	71·0	50·9	68·0	69·4	65·5	70·8
Newport, R. I.	76·2	78·8	79·7	72·4	72·2	77·8	74·6	78·2	74·1	65·4	81·6	77·1	75·7
New Shoreham, R. I.	.....	.....	*u* 80·6	73·6	67·8	71·3	71·4	75·2	76·7	65·2	82·6	81·2	.....
New York city, N. Y.	69·1	72·8	71·3	69·2	69·6	77·7	78·2	77·8	75·1	62·9	77·0	73·9	72·9
Norfolk, Va.	67·8	73·8	70·4	66·8	73·7	74·8	78·3	70·8	63·1	66·2	71·7	68·9	70·5
North Platte, Neb.	61·6	58·5	60·7	57·7	63·0	70·0	69·8	73·7	70·3	57·9	65·0	66·2	64·6
Olympia, Wash.	68·8	72·3	77·6	81·7	*x* 86·3	88·5	88·8	87·1	81·8	77·5	68·3	72·8	78·9
Omaha, Neb.	61·4	65·3	66·5	63·8	68·3	73·2	77·1	79·4	74·6	69·6	70·3	65·8	69·6
Oswego, N. Y.	68·8	70·2	69·7	67·4	72·0	73·5	74·6	74·3	76·3	64·6	74·5	71·8	71·5
Pembina, Dak.	73·6	*w* 76·0	.....	.....	.....	.....	.....	.....	.....	.....	.....	.....	.....
Pensacola, Fla.	75·4	76·1	77·0	75·2	76·2	72·1	78·8	72·5	64·9	77·8	70·3	74·5	74·2
Philadelphia, Pa.	65·2	70·2	67·0	67·5	67·9	74·3	77·1	74·3	72·8	61·8	70·6	70·4	69·9
Pike's Peak, Col.	71·4	73·6	65·2	67·3	60·6	61·0	60·8	62·3	65·7	70·8	65·4	48·3	65·1
Pioche, Nev.	19·1	23·1	27·6	30·2	36·1	66·8	54·4	43·0	58·1	82·5	25·1	13·0	34·1
Pittsburg, Pa.	66·8	70·0	70·2	70·4	75·2	78·1	80·4	71·2	69·7	63·5	62·9	69·0	70·6
Port Huron, Mich.	73·4	75·3	73·1	74·6	79·0	80·0	80·5	79·8	81·9	72·7	74·5	76·3	76·8
Portland, Me.	67·7	66·5	71·5	66·5	65·1	70·2	67·3	68·9	69·5	58·9	71·6	66·5	67·1
Portland, Ore.	61·8	68·7	70·6	81·7	74·4	79·3	78·2	81·5	73·8	67·5	54·0	66·4	71·5
Portsmouth, N. C.	*x* 79·3	*y* 81·5	*z* 87·8	A 84·2	B 83·3	C 75·1	*x* 83·9	73·7	80·1	82·4	78·5	81·6	81·0
Prescott, Ariz.	41·3	46·6	40·8	38·0	42·1	63·8	47·3	38·1	43·7	36·4	24·4	18·0	40·0
Punta Rassa, Fla.	71·4	75·3	73·2	72·6	79·9	72·9	78·2	71·0	67·6	70·4	68·8	70·7	72·7
Red Bluff, Cal.	31·2	D 32·5	30·7	36·9	43·8	87·2	73·4	79·3	60·3	66·5	44·1	40·3	52·7
Rochester, N. Y.	61·8	65·4	65·8	70·5	74·5	80·2	79·6	77·6	80·0	62·2	70·0	68·5	71·3
Roseburg, Ore.	57·5	61·4	63·8	75·2	82·6	84·0	86·8	82·0	75·7	69·2	57·4	66·5	71·9

Sacramento, Cal.	55·9	56·1	54·9	54·3	51·6	87·9	81·8	82·2	68·3	70·6	55·8	54·9	64·5
Salt Lake City, Utah	34·4	26·2	28·7	31·0	37·5	55·3	48·0	51·9	39·6	41·7	38·8	36·5	37·0
San Antonio, Tex.	65·5	72·4	74·7	68·8	70·2	73·7	67·4	64·3	51·5	63·8	71·8	58·4	66·9
San Diego, Cal.	75·3	77·2	78·4	69·9	59·1	73·3	65·5	72·5	73·0	71·4	68·4	69·2	71·1
Sandusky, Ohio	67·4	70·8	67·8	70·2	70·9	77·0	70·5	72·4	76·0	67·2	66·6	60·9	71·1
Sandy Hook, N. J	72·0	77·3	72·2	66·8	65·7	69·5	73·2	75·8	76·7	68·4	70·4	76·6	72·9
San Francisco, Cal	81·7	81·5	79·3	71·1	57·8	84·8	72·4	80·2	69·8	77·4	74·9	76·7	75·6
Santa Fé, N. M	40·2	40·2	46·6	48·0	55·2	55·7	60·3	56·8	48·7	38·7	35·8	30·2	45·8
Savannah, Ga	69·1	75·5	76·3	76·2	79·2	69·4	75·8	63·9	56·0	62·4	60·5	63·2	69·7
Shreveport, La	73·4	71·9	78·9	70·9	69·2	71·3	71·0	73·0	61·5	62·9	74·6	64·9	70·2
Smithville, N. C	71·1	78·5	76·1	74·7	81·1	77·7	82·8	78·4	73·2	85·0	81·3	76·2	78·0
Springfield, Ill	61·7	59·6	61·9	58·7	63·2	69·1	67·7	70·6	70·9	67·1	64·2	71·5	65·5
Springfield, Mass.	70·9	74·1	73·4	70·9	65·0	63·7	68·7	65·1	68·2	68·1	72·6	68·1	68·2
St. Louis, Mo	60·9	60·6	62·7	60·7	70·6	75·2	77·0	78·2	71·5	63·8	66·1	69·0	68·2
St. Paul, Minn	71·0	68·6	64·6	63·0	69·7	74·0	72·8	72·5	72·1	60·5	64·6	68·0	68·5
St. Michael's, Alaska	86·8	88·6	90·6	95·7	96·1	98·8	99·3	100·0	97·6	98·6	97·8	88·3	94·4
Sitka, Alaska	......	......	......	......	......	......	......	......	......	E 72·8	80·3	80·6	......
St. Vincent, Minn.	......	......	J 75·7	76·5	84·5	91·2	98·2	90·9	84·6	75·6	69·2	74·3	......
Socorro, N. M	......	......	......	......	......	......	k 52·3	47·1	F 33·1	......	......	......	......
Thatcher's Island, Mass	87·3	87·2	87·7	80·6	74·4	72·9	70·0	74·2	80·9	79·8	88·9	89·5	81·1
Toledo, Ohio	68·9	74·1	68·9	68·5	69·7	77·2	68·9	79·4	76·4	69·6	68·5	71·8	72·6
Umatilla, Ore	40·6	43·8	48·6	54·4	76·2	82·5	82·5	82·2	64·3	55·5	48·5	48·3	60·2
Vicksburg, Miss	72·0	75·8	81·3	75·8	77·8	74·9	72·2	68·6	57·1	58·6	63·7	68·2	70·1
Virginia City, Mont.	36·9	36·9	31·4	48·7	G 62·4	......	......	......	......	......	......	......	......
Visalia, Cal.	41·6	38·1	41·1	54·7	58·8	81·3	80·3	77·1	66·4	64·7	45·7	40·8	57·5
Washington, D. C	65·1	70·8	68·0	67·9	71·2	74·2	77·6	73·4	67·3	66·0	69·3	72·8	70·4
Wilmington, N. C.	73·4	79·3	72·3	71·9	80·5	74·4	74·7	66·4	50·2	62·6	71·1	72·2	71·5
Winnemucca, Nev.	17·8	16·6	18·7	32·9	46·7	H 77·7	64·1	62·8	49·4	40·2	32·6	27·0	40·5
Wood's Holl, Mass.	81·9	83·0	80·8	71·5	70·5	74·2	74·6	76·3	75·8	69·3	81·8	81·5	76·8
Yankton, Dak.	61·9	60·5	58·8	62·1	65·1	69·2	69·8	76·2	77·8	72·9	65·0	68·2	67·4
Yuma, Ariz.	......	......	......	......	......	......	......	f 40·0	41·0	43·0	38·7	36·4	......

Notes to Table on Pages 210–213.

a Local observations discontinued Oct. 13, 1880.
b For 15 days only.
c Local observations discontinued Dec. 21, 1880.
d For 9 days only.
e Station closed November 30, 1880.
f Observations discontinued Dec. 31, 1880.
g Observations commenced Oct. 13, 1880.
h For 30 days only.
i Office burned Dec. 23, 1880; no observations taken Dec. 23, 24, and 25.
j Observations commenced Dec. 3, 1880.
k Local observations commenced Jan. 1, 1881.
l Local observations discontinued March 31, 1881.
m Observations commenced Jan. 3, 1881.
n Observations commenced Dec. 1, 1880.
o For 28 days only.
p For 29 days only.
q Local observations discontinued July 31, 1880.
r For 8 days only; office destroyed by fire on Nov. 17, 1880.
s Observations commenced Jan. 1, 1881.
t For 30 days only.
u Observations commenced Sept. 1, 1880.
v For 29 days only.
w Station closed Sept. 3, 1880.
x For 25 days only.
y For 26 days only.
z For 24 days only.
A For 18 days only.
B For 21 days only.
C For 23 days only.
D August 16 to 31 only.
E Observations commenced April 1, 1881.
F Local observations discontinued March 19, 1881.
G Office removed to Eagle Rock, Idaho, on Nov. 19, 1880.
H For 3 days only; office and records destroyed by fire, Dec. 28, 1880.
I Local observations commenced Feb. 1, 1881.
J Observations commenced Sept. 4, 1880.

Notes to Table on Pages 215–219.

a Too small to measure.
b Station closed Nov. 30, 1880.
c Station closed Dec. 4, 1880.
d Observations discontinued Dec. 31, 1881.
e Observations commenced Oct. 13, 1880.
f Station closed July 27, 1881.
g Observations commenced Dec. 3, 1880.
h For 14 days only.
i For 20 days only.
j Station closed Feb. 12, 1881.
k No mail reports received prior to Aug. 1, 1881.
l Data incomplete; observer sick.
m Observations commenced Sept. 27, 1880.
n May 20 to 31 only.
o Observations commenced Jan. 3, 1881
p No rain-gauge at this station during prior months.
q Observations commenced Dec. 1, 1880.
r Record for October incomplete.
s For 24 days only.
t For 25 days only.
u Office destroyed by fire Nov. 17, 1880, no record of rainfall during month.
v Observations commenced Jan. 1, 1881.
w Observations commenced Sept. 1, 1880.
x Station closed Sept. 3, 1880.
y Station closed.
z Observations commenced April 10, 1881.
A August 16 to 31 only.
B Observations commenced March 30, 1881.
C Record of rainfall for August incomplete.
D Station closed May 23, 1881.
E Observations commenced Feb. 1, 1881.
F Observations commenced Sept. 4, 1880.
G Office moved to Eagle Rock, Idaho, on Nov. 19, 1880.

Monthly and Annual Amounts of Precipitation (in Inches and Hundredths), from July, 1880, to June, 1881, inclusive.—(United States Signal Service.)

STATION.	Elevation of rain-gauge above ground, feet.	1880. July.	August.	September.	October.	November	December.	1881. January.	February.	March.	April.	May.	June.	Annual amount.
Albany, N. Y.	70	3·78	2·84	2·86	2·43	2·40	2·01	2·86	2·50	3·80	1·34	3·00	3·76	34·59
Alpena, Mich.	52	6·14	1·61	3·31	1·05	2·49	1·85	1·59	5·04	3·04	·88	5·02	3·23	36·15
Atlanta, Ga.	92	3·16	3·61	6·21	2·81	8·21	5·70	8·35	10·41	10·08	4·58	1·27	2·46	67·75
Atlantic City, N. J.	37	5·80	4·42	3·14	4·02	4·03	8·30	7·18	5·08	4·71	2·85	1·08	3·82	55·48
Augusta, Ga.	40	5·98	5·10	1·04	2·08	5·31	4·13	8·09	3·98	7·54	4·71	1·35	2·11	51·97
Baltimore, Md.	60	6·47	4·44	1·78	2·64	2·86	4·89	4·84	5·68	7·59	2·00	2·30	7·81	53·30
Barnegat, N. J.	8	5·78	6·15	3·14	4·05	4·35	5·95	8·43	5·39	7·58	1·36	1·77	6·18	60·13
Bismarck, Dak.	1	2·02	4·88	·72	·27	·87	1·17	·09	·78	·46	1·02	2·27	4·11	19·19
Boisé City, Idaho.	32	·08	·02	·11	·50	·48	4·01	3·68	3·51	·64	1·34	·07	·39	14·01
Boston, Mass.	103	6·86	2·90	2·36	3·15	2·30	3·42	7·21	4·80	9·86	1·66	4·16	7·79	56·56
Brackettville, Tex.	4	5·35	4·39	1·77	·27	1·63	1·33	·35	·30	·81	3·43	5·29	·01	24·92
Breckinridge, Minn.	2	2·52	3·67	·08	1·78	·73	(b)	—	—	—	—	—	—	—
Brownsville, Tex.	40	3·64	16·58	1·90	3·82	3·44	·58	2·73	1·18	·20	·30	3·43	(a)	37·80
Buffalo, N. Y.	67	3·31	6·07	2·96	4·87	3·17	3·03	2·31	2·43	3·74	·76	2·18	4·15	38·98
Burkes, Ariz.	12	·04	·48	·08	(a)	·00	(c)	—	—	—	—	—	—	—
Burlington, Vt.	72	2·30	2·26	3·20	6·22	2·57	·63	·88	1·79	1·56	·62	2·27	1·89	26·24
Cairo, Ill.	78	4·34	2·61	4·55	6·98	3·98	2·22	3·56	4·07	1·33	3·02	2·44	1·81	42·36
Campo, Cal.	2	·12	·41	·01	·68	·85	4·85	1·74	·53	5·00	1·52	·12	·04	15·87
Camp Thomas, Ariz.	—	·87	2·49	·55	·18	·03	1·27	·03	·18	1·21	·03	·07	·00	7·46
Cape Hatteras, N. C.	1	9·06	16·30	3·65	12·00	12·08	(b)	—	—	—	—	—	—	—
Cape Henry, Va.	14	6·27	9·86	4·17	3·61	8·11	7·10	4·08	4·06	4·00	6·03	1·34	3·93	62·67
Cape Lookout, N. C.	1	7·11	4·43	5·04	7·00	7·61	d 2·11	—	—	—	—	—	—	—
Cape May, N. J.	6	4·58	7·49	2·91	5·07	4·79	9·21	7·36	4·28	4·04	1·34	3·51	5·11	60·54
Castroville, Tex.	8	7·74	6·69	3·95	1·34	1·76	·20	·45	·85	·58	3·06	3·09	(a)	30·40
Cedar Keys, Fla.	35	9·10	19·45	4·41	10·37	6·29	5·05	9·36	3·51	8·86	3·45	2·25	1·09	78·79
Champaign, Ill.	—	—	—	—	e ·04	1·80	1·53	·89	6·49	5·30	1·36	3·63	4·71	—
Charleston, S. C.	57	5·77	3·07	4·89	9·19	5·50	3·41	5·98	1·56	4·11	3·37	·48	1·47	48·80
Charlotte, N. C.	47	5·62	10·37	1·84	2·94	5·96	6·42	6·81	3·95	3·77	3·01	2·26	1·35	55·10
Chattanooga, Tenn.	59	5·13	3·33	5·20	1·09	8·82	5·46	4·70	6·19	6·85	4·56	4·46	5·24	61·63
Cheyenne, Wyo.	24	1·88	2·23	1·05	·76	·36	·08	·36	·22	·32	2·32	1·14	1·22	11·94
Chicago, Ill.	92	3·07	4·47	2·25	3·09	·87	1·01	·87	5·98	2·99	1·84	1·85	5·93	34·28
Chincoteague, Va.	30	4·67	8·32	3·36	3·58	4·50	7·01	4·04	4·72	4·22	4·33	1·73	2·34	52·82

Monthly and Annual Amounts of Precipitation, etc., from July, 1880, to June, 1881, inclusive.—(Continued.)

STATION.	Elevation of rain-gauge above-ground, feet.	1880. July.	August.	September.	October.	November	December.	1881. January.	February.	March.	April.	May.	June.	Annual amount
Cincinnati, Ohio	76	2·46	4·01	1·37	2·98	4·42	4·26	3·76	4·05	3·51	3·25	2·23	7·82	45·02
Cleveland, Ohio	74	4·75	3·97	2·98	2·24	2·27	1·51	1·96	2·57	2·82	1·75	·74	8·07	35·63
Coleman City, Tex.	0	3·13	1·78	7·05	1·78	1·46	1·90	·36	·61	·55	1·65	2·70	*f* ·02	22·99
Columbus, Ohio	70	4·86	6·95	1·80	2·35	4·54	3·98	2·25	4·44	4·01	2·04	2·00	4·02	43·24
Concho, Tex.	0	10·28	2·25	7·45	2·63	1·33	·70	·22	·51	·28	1·87	6·82	·27	34·58
Corsicana, Tex.	30	3·43	·58	7·75	1·02	5·54	·64	2·20	3·37	2·71	3·73	14·33	·00	46·20
Davenport, Ia.	77	4·31	5·90	4·87	·94	1·23	1·15	1·34	4·14	3·33	1·11	1·34	7·04	37·60
Dayton, Wash.	2	1·68	1·29	·19	1·05	2·00	7·93	5·08	5·04	1·84	3·51	4·55	1·61	25·22
Deadwood, Dak.	24	1·51	3·33	·30	1·27	2·37	·04	3·10	1·26	2·50	2·05	3·70	3·04	25·16
Decatur, Tex.	5	3·70	·14	9·13	1·65	1·31	·62	·08	3·78	1·81	2·98	5·82	·01	31·97
Del. Breakwater, Del.	27	5·30	5·06	1·23	3·84	2·61	3·29	4·30	2·92	2·94	1·08	1·23	3·44	37·33
Denison, Tex.	28	5·89	2·14	6·74	2·53	3·21	·82	·74	4·85	1·02	3·31	8·03	·00	39·88
Denver, Col.	86	1·38	1·46	·80	1·37	·83	·10	·49	1·22	·87	·50	2·21	·09	11·41
Des Moines, Ia.	45	3·82	6·69	5·34	4·90	1·97	·86	1·55	2·68	1·78	3·36	3·82	15·79	52·56
Detroit, Mich.	71	5·74	5·51	4·26	5·45	3·02	1·22	2·39	6·41	3·36	2·37	2·45	5·90	48·08
Dodge City, Kan.	30	4·00	5·17	·32	1·42	2·43	·03	·15	1·03	·50	2·38	12·82	1·77	32·02
Dubuque, Ia.	43	3·55	7·15	6·84	·66	2·11	1·25	1·87	3·75	3·78	1·30	2·20	7·56	42·02
Duluth, Minn	28	1·98	4·35	1·86	2·32	2·80	1·02	1·65	1·79	1·46	1·10	1·77	2·52	25·31
Eagle Pass, Tex	2	6·68	5·78	7·16	2·70	·55	·30	1·15	·34	·32	·87	8·16	·00	34·01
Eagle Rock, Idaho	1	......	......	......	......	......	*g* 4·50	3·66	5·14	·07	3·13	·74	1·09	......
Eastport, Me.	50	3·45	1·48	4·28	5·11	5·07	3·53	1·64	4·11	6·78	1·06	18·22	4·47	54·80
Edinburg, Tex.	28	1·88	6·23	*h* 2·38	*i* 2·59	1·53	·11	2·79	*j* 1·21	......	......	......	......	......
El Paso, Tex.	14	6·54	3·60	·80	·47	·02	1·58	·35	·24	·01	·22	1·83	·02	15·63
Erie, Pa.	61	3·06	4·04	5·01	3·56	4·45	1·25	2·10	1·93	2·30	1·26	2·16	6·37	37·58
Escanaba, Mich.	38	2·80	3·38	2·84	1·14	2·39	2·23	1·91	2·10	1·00	·44	7·91	6·21	34·44
Florence, Ariz.	4	1·22	·54	·55	·13	·00	1·20	·00	·03	1·98	·86	·13	·00	6·64
Ft. Apache, Ariz.	0	5·83	1·41	·55	·56	·03	2·38	·20	1·17	2·45	1·53	·35	(*a*)	16·49
Ft. Assiniboine, Mont.	......	......	*k* 2·56	·37	·22	·60	1·61	2·29	·70	·38	·21	(*l*)	(*l*)	......
Ft. Bennett, Dak.	17	......	......	*m* ·00	·15	·08	·55	·82	1·18	1·57	·93	4·09	3·07	15·21
Ft. Benton, Mont.	58	1·12	1·56	·32	1·09	1·44	1·39	2·27	·66	·29	·18	*n* 1·43	3·46	......
Ft. Buford, Dak.	0	4·17	2.36	1·04	·00	·66	3·08	1·98	1·10	1·17	1·34	1·00	3·44	22·24
Ft. Custer, Mont.	82	2·51	2·55	·20	1·19	·54	1·27	2·84	·51	·17	·78	1·80	1·57	15·93
Ft. Davis, Tex.	2	10·11	5·57	3·16	·38	·56	·31	·50	·14	·27	1·04	6·31	·07	28·42

Ft. Elliott, Tex.	0	2·11	1·70	·54	2·40	·10	·38	·47	·74	(a)	1·36	5·27	·10	15·04
Ft. Gibson, Ind. Ter.	35	3·61	2·82	1·89	3·23	2·34	·08	1·76	4·14	2·38	2·04	6·06	3·80	36·14
Ft. Grant, Ariz.	1	5·63	3·37	1·01	·47	·00	1·57	·05	·33	·89	·84	·26	(a)	14·42
Ft. Griffin, Tex	3	3·54	·40	2·43	1·11	2·13	1·20	·82	1·30	·17	3·71	3·70	·09	20·49
Ft. Keogh, Mont.	37	2·06	2·10	·07	·29	·56	1·12	2·85	·36	·31	·35	·76	4·44	15·37
Ft. Macon, N. C.	5	.....	.....	.....	.....	.....	.....	o 5·66	4·09	5·29	4·56	1·74	7·71	.....
Ft. McKavett, Tex.	21	6·78	2·41	5·20	2·34	1·61	·49	·12	·19	·51	2·27	5·26	·07	27·25
Ft. Missoula, Mont.	.....	.....	.....	.....	.....	.....	p 4·43	3·22	2·40	·47	1·38	·29	2·75	.....
Ft. St. Michaels, Alaska	1	·65	4·05	5·61	2·38	1·48	1·31	·06	·00	·01	·39	2·66	·65	19·25
Ft. Shaw, Mont.	24	1·51	1·11	·05	·25	·89	1·17	2·50	·40	·15	·35	1·46	3·76	18·09
Ft. Sill, Ind. T.	2	5·57	3·06	1·17	·54	·60	·63	1·92	3·45	1·28	2·63	6·04	·21	28·00
Ft. Stevenson, Dak.	4	·83	1·96	1·08	·36	·88	·91	·72	·86	·14	·60	2·81	5·88	16·18
Ft. Verde, Ariz	4	1·86	·97	·10	·57	·13	1·56	·07	·12	2·64	·97	·07	(a)	9·14
Fredericksburg, Tex.	24	5·40	4·55	12·25	2·69	5·09	1·32	·95	·59	1·13	2·13	6·30	·00	42·40
Galveston, Tex.	52	2·48	1·62	10.20	2·20	8·85	2·16	3·94	3·29	1·47	4·76	3·50	·03	49·50
Grand Haven, Mich.	76	4·47	3·08	3·74	3·15	2·01	2·32	2·34	6·13	3·77	1·89	1·19	3·44	42·38
Hatteras, N. C.	1	.....	.....	.....	.....	.....	q 4·46	7·27	5·87	6·04	6·88	1·13	4·99	.....
Helena, Mont.	20	·86	1·38	·00	1·33	·87	4·64	2·86	·51	·00	1·55	1·00	3·51	19·01
Henrietta, Tex.	4	3·77	1·82	3·53	(r)	1·20	·51	·99	·86	·98	4·87	3·00	·88	.....
Indianapolis, Ind.	74	2·26	2·67	1·86	3·54	2·58	1·44	2·10	0·43	4·01	2·60	3·78	3·92	37·19
Indianola, Tex.	40	4·83	7·45	8·21	3·43	4·18	·69	3·07	4·21	·29	3·29	2·02	·00	41·06
Jacksboro', Tex	18	.....	s ·97	18·95	t 4·97	1·58	·51	·48	2·08	2·23	2·36	3·53	·04	.....
Jacksonville, Fla.	57	5·04	8·96	5·21	16·25	6·09	1·29	9·12	1·12	2·89	4·57	2·61	2·82	66·87
Keokuk, Ia.	60	2·26	3·81	3·21	2·08	1·13	·67	·50	2·58	2·42	3·12	1·35	3·70	31·76
Key West, Fla	52	4·61	7·30	3·59	2·81	1·95	·71	3·65	·35	·46	4·09	4·01	7·47	41·90
Kittyhawk, N. C.	1	9·94	11·18	5·03	5·95	9·92	5·74	7·34	3·97	4·24	4·97	·80	5·06	74·14
Knoxville, Tenn.	77	6·28	4·45	3·64	·71	0·46	3·74	4·41	4·20	3·90	2·09	1·14	5·13	47·05
La Crosse, Wis.	67	3·22	4·42	3·22	1·37	2·40	·71	1·51	1·26	1·01	1·35	3·13	2·75	26·36
La Mesilla, N. M.	16	1·02	1·08	1·02	·35	·13	1·11	·04	(a)	·14	·09	1·75	·43	8·01
Laredo, Tex.	5	5·81	8·08	1·05	·72	1·07	·58	3·15	·31	·27	2·53	7·75	·00	31·27
Leavenworth, Kan.	48	6·86	7·06	2·78	3·09	2·40	·40	·44	4·84	2·21	1·86	3·65	5·27	41·46
Lewiston, Idaho	38	1·86	1·09	·20	1·54	2·36	6·31	4·46	4·33	·49	2·00	·28	2·30	27·74
Little Rock, Ark.	57	3·30	5·53	5·00	2·10	6·84	3·08	2·07	6·34	2·38	1·94	5·51	6·08	50·06
Los Angeles, Cal.	50	(a)	(a)	·00	·14	·67	8·40	1·43	·36	1·06	·46	·01	·00	13·13
Louisville, Ky.	102	2·18	2·82	2·44	6·39	3·86	3·52	3·24	4·05	2·46	3·46	2·35	3·78	40·51
Lynchburg, Va.	50	5·75	3·20	1·57	1·33	3·26	1·63	4·47	2·37	2·57	2·14	2·20	1·67	38·16
Madison, Wis.	57	6·00	5·90	4·44	1·08	1·68	1·17	2·05	5·42	4·36	1·50	4·25	4·15	42·60
Mason, Tex.	2	5·49	4·22	10·39	2·16	2·14	·78	·87	·70	·85	2·95	5·29	·00	35·39
Marquette, Mich.	57	3·72	3·70	2·83	2·36	1·48	1·27	1·69	2·80	3·84	·52	4·47	3·47	32·24
Memphis, Tenn.	51	2·14	3·03	2·47	5·20	8·39	2·62	4·38	6·41	3·23	5·74	2·80	2·83	49·24
Milwaukee, Wis.	135	3·20	3·54	1·85	·51	·82	·74	1·86	5·39	5·30	·78	2·55	3·77	30·31
Mobile, Ala.	83	4·02	4·75	7·04	7·82	(u)	3·71	7·62	8·00	10·41	9·21	1·44	4·85	.....

Monthly and Annual Amounts of Precipitation, etc., from July, 1880, to June, 1881, inclusive.—(Continued.)

STATION.	Elevation of rain-gauge above-ground, feet.	1880. July.	1880. August.	1880. September.	1880. October.	1880. November	1880. December.	1881. January.	1881. February.	1881. March.	1881. April.	1881. May.	1881. June.	Annual amount
Montgomery, Ala	58	3·17	4·41	2·83	2·06	4·06	5·68	3·58	7·05	5·45	4·52	1·41	3·04	47·86
Moorhead, Minn	41							v ·60	2·13	·02	·77	2·80	5·59	......
Morgantown, W. Va	1	3·80	7·25	3·18	3·07	1·90	3·99	3·08	3·56	2·12	2·30	3·28	4·77	42·80
Mt. Washington, N. H.	36	7·24	5·82	15·23	7·96	9·37	7·80	3·94	6·62	8·51	5·08	12·50	7·03	97·10
Nashville, Tenn	49	5·69	2·22	5·39	7·24	5·77	3·32	3·54	5·48	2·70	5·12	3·67	3·70	53·03
New Haven, Conn	108	4·90	8·14	3·73	4·07	2·82	3·49	4·79	6·17	10·42	1·71	3·89	5·14	59·27
New London, Conn	57	5·59	6·53	3·06	4·14	2·05	4·03	6·76	7·06	8·07	2·23	4·85	3·75	58·12
New Orleans, La	77	11·22	4·60	7·48	1·88	6·04	6·45	11·15	5·80	2·75	3·02	3·20	2·84	67·33
Newport, R. I	43	5·86	7·15	2·80	3·59	4·01	4·87	5·70	8·28	8·24	2·11	2·90	5·94	61·45
New Shoreham, R. I	23			w 4·04	4·02	4·02	4·42	2·16	4·74	5·08	1·91	5·51	12·93	......
New York, N. Y.	145	6·67	4·40	2·26	2·81	2·40	4·15	5·41	5·06	6·78	1·00	2·38	6·23	49·50
Norfolk, Va	52	7·84	9·90	4·07	3·08	5·64	5·15	3·55	2·88	3·38	4·06	1·49	3·74	54·78
North Platte, Neb	8	2·87	3·96	1·53	2·72	·23	·37	·16	·76	1·26	·87	4·84	6·12	25·69
Olympia, Wash	38	·52	·22	1·05	2·83	8·06	16·66	8·90	16·28	4·03	4·93	1·54	1·03	61·05
Omaha, Neb	75	5·36	7·10	2·01	3·54	·70	·28	·61	3·09	·72	4·23	7·04	5·56	42·04
Oswego, N. Y	62	1·74	3·25	1·51	4·32	6·45	4·24	4·15	4·74	5·08	1·19	2·41	2·98	42·06
Pembina, Dak	18	4·75	4·34	(x)										
Pensacola, Fla	36	7·39	4·68	11·54	7·08	5·86	2·24	7·49	8·99	3·80	4·53	1·40	4·27	69·27
Philadelphia, Pa	95	7·74	5·09	1·10	1·74	1·75	4·05	3·66	4·76	3·83	0·61	2·71	3·87	40·91
Phœnix, Ariz	19	1·18	·72	·67	·20	·00	1·61	·00	·20	1·46	1·10	0·12	·00	7·26
Pike's Peak, Col	1	6·69	4·30	3·87	4·64	4·07	1·60	2·58	1·47	4·44	4·64	3·71	·87	42·97
Pilot Point, Tex	36	2·42	1·38	7·63	1·88	2·16	·93	·75	5·96	2·72	(y)			......
Pioche, Nev	23	·18	·47	·18	·52	·29	1·84	·47	·29	·47	1·08	·21	·03	6·08
Pittsburg, Pa	86	2·15	3·62	3·12	2·80	1·78	2·95	3·55	3·45	3·35	1·81	2·34	6·95	37·82
Port Huron, Mich	63	4·07	4·51	2·17	4·06	2·79	1·23	3·50	4·91	4·00	·76	1·66	3·82	38·98
Port Eads, La	2										z 7·18	·41	·94	......
Portland, Me	77	3·12	2·32	3·20	4·21	3·25	3·17	4·30	5·30	5·09	1·48	5·64	3·09	45·02
Portland, Ore	60	·59	1·31	1·34	1·47	8·17	13·98	8·57	13·36	2·83	3·51	1·36	2·34	53·80
Portsmouth, N. C.	29	3·94	9·56	7·87	5·12	6·17	2·49	2·72	5·98	6·57	4·70	2·13	4·95	62·20
Prescott, Ariz	5	2·34	2·80	1·26	·18	·42	1·84	·16	·10	2·01	·67	·44	(a)	13·12
Punta Rassa, Fla	35	3·76	8·55	2·50	8·10	1·81	1·98	4·02	·52	1·18	·54	3·13	2·53	38·62
Red Bluff, Cal	32	·00	A ·00	·00	·08	·14	12·85	9·40	2·79	·51	1·88	·79	·51	28·90
Rio Grande City, Tex	0	·96	9·13	2·05	1·48	1·15	·35	3·48	1·06	·07	2·12	2·98	(a)	24·78

Rochester, N. Y.	96	3·49	5·53	2·72	5·31	3·19	2·90	4·68	2·60	5·33	1·15	5·86	3·30	46·31
Roseburg, Ore.	82	·01	·41	·47	·77	·77	8·29	11·60	8·19	1·95	1·07	·73	2·92	37·18
Sacramento, Cal	54	(a)	·00	·00	·00	·05	11·81	6·14	5·00	1·37	1·64	(a)	·50	26·57
Salt Lake City, Utah	73	·20	·74	·56	·40	1·17	1·90	1·34	2·44	·88	2·37	2·55	·28	14·73
San Antonio, Tex	83	6·30	8·67	2·93	2·35	2·06	·40	·58	1·24	1·06	3·35	4·70	(a)	33·54
San Diego, Cal	80	·09	·32	·00	·53	·28	4·15	·52	·45	1·88	1·35	·04	·05	9·66
Sandusky, Ohio	66	6·06	5·03	2·02	2·27	2·39	1·06	1·31	3·25	2·75	2·47	·08	10·09	39·08
Sandy Hook, N. J.	1	6·43	4·30	3·35	3·07	3·70	2·51	5·03	5·37	6·92	1·41	3·33	6·80	58·14
San Francisco, Cal	75	·00	·00	·00	·05	·33	12·38	6·09	4·06	·90	2·00	·22	·09	26·86
Santa Fé, N. M	18	2·09	1·79	1·13	·75	·28	·66	·88	·22	·57	·96	2·31	·08	11·84
Savannah, Ga.	58	7·24	3·94	4·65	6·58	4·40	7·48	5·77	1·16	3·70	3·32	·83	·91	50·07
Shreveport, La	44	10·07	1·21	11·61	2·72	7·43	2·24	2·24	6·51	1·80	2·81	6·63	·33	58·55
Silver City, N. M	1	3·05	3·51	3·57	1·66	·00	1·65	·02	·78	·91	·48	1·01	·43	17·02
Sitka, Alaska	96	....	....	....	....	....	....	....	....	....	B 4·21	3·70	1·54	....
Smithville, N. C.	35	5·69	7·49	3·33	6·08	4·77	2·88	6·05	2·31	3·92	4·76	2·25	·82	50·59
Socorro, N. M	13	3·85	(c)	2·07	2·81	·81	3·54	·53	·01	·07	·23	D ·51	....	....
Spokane Falls, Wash	....	....	....	....	....	....	....	....	E 3·86	1·07	1·20	·50	1·23	....
Springfield, Ill	61	1·82	1·80	3·15	2·04	1·00	1·10	·64	5·86	4·45	1·96	2·80	4·96	32·43
Springfield, Mass	64	6·55	3·46	2·65	4·31	3·57	1·82	3·16	4·50	4·40	2·49	4·85	9·07	50·83
St. Paul's Isl'd, Alaska	1	....	....	....	....	....	....	....	....	....	....	....	2·22	....
St. Louis, Mo	100	5·17	1·53	3·10	2·09	2·67	1·80	·49	4·16	1·95	3·14	3·96	2·74	32·80
St. Paul, Minn	58	2·75	3·63	1·74	2·18	2·93	2·09	4·34	2·05	1·06	·47	4·34	2·87	31·65
St. Vincent, Minn	14	....	....	F 2·19	2·69	·07	·73	·13	·56	·08	·81	4·00	3·47	....
Stockton, Tex	2	7·27	6·86	13·30	·01	1·70	·34	·20	·13	·06	·46	2·26	1·05	34·60
Thatcher's Isl'd, Mass	4	6·51	4·15	3·72	2·81	4·32	2·81	4·56	4·73	9·96	1·56	7·22	7·51	59·86
Toledo, Ohio	105	6·22	4·65	1·23	3·31	1·74	·60	·54	4·28	1·90	1·76	·45	7·36	34·13
Tucson, Ariz	16	1·62	1·38	1·89	·09	·00	·57	·08	·35	1·17	·02	·04	·00	7·53
Umatilla, Ore	3	·48	1·14	·13	·35	·54	3·05	2·45	1·92	·44	·80	·06	·96	13·06
Uvalde, Tex	0	4·82	5·71	4·46	·44	1·64	1·23	·43	·82	·50	2·39	6·44	·00	28·97
Vicksburg, Miss	53	4·89	5·67	10·51	5·75	14·15	4·10	3·37	7·30	3·53	1·48	4·39	1·94	66·98
Virginia City, Mont	29	2·06	·82	·14	2·40	G 1·20	....	....	....	....	....	....	....	....
Visalia, Cal	44	(a)	·00	·00	·13	·35	5·08	2·71	1·10	1·20	·86	·29	·00	11·67
Washington, D. C.	51	2·25	3·83	3·43	2·31	2·48	4·02	5·14	4·01	6·61	2·08	1·86	5·71	43·72
Wickenburg, Ariz	2	2·20	·01	·89	·08	·00	1·23	·07	·00	1·70	·70	·07	(a)	7·64
Wilmington, N. C.	45	9·20	8·70	1·20	5·23	5·30	1·52	5·06	2·80	5·14	3·47	2·11	3·48	53·35
Winnemucca, Nev	5	·00	·02	·00	·00	·10	1·88	3·08	1·59	·87	1·04	1·08	·02	10·22
Wood's Holl, Mass	34	3·77	5·39	1·27	3·18	1·09	3·03	2·57	5·46	4·22	2·12	2·30	6·25	41·31
Yankton, Dak	28	3·32	3·04	·98	1·08	·21	·99	1·13	2·70	1·74	2·49	9·88	3·51	31·07
Yuma, Ariz	26	(a)	(a)	(a)	(a)	·00	·74	·00	(a)	(a)	·55	·00	(a)	1·29

Elevations of Signal-Service Barometers above Mean Sea-Level on June 30, 1881, and of Thermometers and Rain-Gauges above-ground (United States Signal Service).

STATION.	AB'VE SEA-LEVEL.	ABOVE-GROUND.	
	Barometer.	Thermometer.	Rain-gauge.
	Feet.	Feet.	Feet.
Albany, N. Y.	75·3	50·9	69·7
Alpena, Mich.	609·4	54·4	52·0
Atlanta, Ga	1,131·3	77·7	92·2
Atlantic City, N. J	12·9	9·7	37·2
Augusta, Ga.	182·8	18·1	39·8
Baltimore, Md.	45·2	33·1	69·1
Barnegat, N. J.	20·0	5·5	7·6
Benton, Mont.	....	50·5	58·0
Bismarck, Dak.	1,704·3	16·4	·5
Boerne, Tex.	1,508·0 B.	6·5	4·5
Boisé City, Idaho	2,768·0 B.	19·3	32·1
Boston, Mass.	142·2	155·9	161·6
Brackettsville, Tex.	1,137·0 B.	6·4	3·5
Brownsville, Tex.	43·4 ?	20·2	40·4
Buffalo, N. Y.	664·5	72·0	67·2
Burlington, Vt.	268·0	54·6	72·0
Cairo, Ill.	377·3	44·4	77·6
Campo, Cal.	2,527·0	5·0	2·1
Cape Hatteras, N. C.	8·4	7·0	1·0
Cape Henry, Va.	16·0	16·5	14·3
Cape Lookout, N. C	15·0	18·0	1·0
Cape May, N. J.	27·0	18·6	6·1
Castroville, Tex.	778·0 B.	16·0	8·0
Cedar Keys, Fla.	22·5	20·3	34·7
Charleston, S. C.	52·5	40 5	56·6
Charlotte, N. C	837·8 ?	34·6	47·0
Chattanooga, Tenn	783·2	42·6	59·2
Cheyenne, Wyo.	6,089·0	15·3	24·0
Chicago, Ill.	660·9	69·2	92·0
Chincoteague, Va	18·5	28·9	29·7
Cincinnati, Ohio.	620·4	67·8	76·3
Cleveland, Ohio	689·7	78·5	74·0
Coleman City, Tex.	1,735·0 B.	4·1	....
Columbus, Ohio	804·6	52·0	70·0
Concho, Tex.	1,888·0 B.	5·6	On ground.
Corsicana, Tex.	447·5	18·6	30·0
Davenport, Iowa	614·7	46·1	77·0
Dayton, Wash.	1,700·0 B.	5·8	1·7
Deadwood, Dak.	4,630·0 B.	15·7	23·9
Decatur, Tex.	1,160·0 ?	17·0	5·0
Delaware Breakwater, Del.	20·0	12·8	26·6
Denison, Tex.	767·4	16·8	28·2
Denver, Col.	5,293·6	45·3	56·1
Des Moines, Iowa	849·0	35·0	45·3

Elevation of Signal Barometers above Mean Sea-Level, etc.—(Continued.)

STATION.	AB'VE SEA-LEVEL	ABOVE-GROUND.	
	Barometer.	Thermometer.	Rain-gauge.
	Feet.	Feet.	Feet.
Detroit, Mich	661.4	61·4	71·1
Dodge City, Kan	2,512·5	15·3	29·9
Dubuque, Iowa	665·1	31·9	43·1
Duluth, Minn	644·1	18·9	27·7
Eagle Pass, Tex	800·0 B.	5·3	·1
Eagle Rock, Idaho	4,780·6	12·2	1·0
Eastport, Me	61·2	32·5	55·5
El Paso, Tex	3,956·0 B.	16·8	14·1
Erie, Pa	681·1	31·6	60·6
Escanaba, Mich	611·6	24·9	38·2
Florence, Ariz	1,553·0 B.	4·9	3 9
Fort Apache, Ariz	5,004·0	7·0	·1
Fort Assiniboine, Mont	....	4·9	Not up.
Fort Bennett, Dak	....	12·0	17·0
Fort Buford, Dak	1,876·0 B.	7·8	....
Fort Custer, Mont	3,100·0 B.	....	....
Fort Davis, Tex	4,918·0 B.	3·2	2·0
Fort Elliott, Tex	....	6·4	On ground.
Fort Gibson, Ind. T	510·1	19·1	35·4
Fort Grant, Ariz	4,737·0 B.	5·6	10·0
Fort Griffin, Tex	1,243·0 B.	7·0	3·0
Fort Keogh, Mont	....	13·8	37·0
Fort Macon, N. C	11·0	8·1	4·8
Fort McKavett, Tex	....	4·1	21·1
Fort Missoula, Mont	....	6·9	....
Fort Shaw, Mont	....	....	....
Fort Sill, Ind. T	1,190·0 B.	5·5	2·0
Fort Stevenson, Mont	1,734·0 B.	7·5	4·5
Fort Verde, Ariz	3,105·0 B.	5·5	3·6
Fredericksburg, Tex	1,742·0 B.	15·8	24·0
Galveston, Tex	39·5	36·1	51·5
Grand Haven, Mich	620·2	22·6	75·8
Hatteras, N. C	19·5	6·3	1·0
Helena, Mont	4,315·6 B.	....	....
Henrietta, Tex	915·0 B.	4·0	3·5
Huron, Dak	1,300·0 ?	....	....
Indianapolis, Ind	753·3	52·2	73·5
Indianola, Tex	25·9	29·2	39·9
Jacksboro, Tex	1,133·0 B.	5·8	17·5
Jacksonville, Fla	43·0	37·4	57·0
Keokuk, Iowa	617·6	46·9	59·5
Key West, Fla	26·9	42·9	52·1
Kittyhawk, N. C	22·0	3·9	1·0
Knoxville, Tenn	980·0	72·4	77·4
La Crosse, Wis	708·0	40·0	67·0

Elevation of Signal Barometers above Mean Sea-Level, etc.—(Continued.)

STATION.	AB'VE SEA-LEVEL	ABOVE-GROUND.	
	Barometer.	Thermometer.	Rain-gauge.
	Feet.	Feet.	Feet.
La Mesilla, N. Mex.	4,124·0 B.	17·8	16·0
Laredo, Tex.	401·0 B.	4·2	5·2
Leavenworth, Kan.	841·9	34·5	48·0
Lewiston, Idaho.	619·0 ?	22·5	37·6
Little Rock, Ark.	298·2	25·6	57·2
Los Angeles, Cal.	350·1	36·6	50·0
Louisville, Ky.	530·0	89·3	102·5
Lynchburg, Va.	651·5	30·5	50·0
Madison, Wis.	949·2	32·6	56·8
Marquette, Mich.	672·9	36·4	56·7
Mason, Tex.	1,620·0 B.	16·3	1·7
Memphis, Tenn.	320·8	52·8	51·0
Milwaukee, Wis.	697·1	105·4	134·8
Mobile, Ala.	68·9	64·5	84·6
Montgomery, Ala.	219·0	33·6	58·2
Moorhead, Minn.	923·0	23·3	41·4
Morgantown, W. Va.	962·6	10·2	1·0
Mount Washington, N. H.	6,259·0 ?	6·0	2·0
Nashville, Tenn.	507·0	34·1	49·0
New Haven, Conn.	106·4	112·4	108·3
New London, Conn.	46·6	28·6	57·2
New Orleans, La.	52·4	45·3	77·1
Newport, R. I.	44·4	19·1	43·0
New River Inlet, N. C.	....	58·0	....
New Shoreham, R. I.	27·4	8·2	22·8
New York, N. Y.	164·3	147·8	144·9
Norfolk, Va.	30·1	20·0	52·5
North Platte, Neb.	2,841·0	18·8	7·5
Olympia, Wash.	36·0	22·9	38·2
Omaha, Neb.	1,113·3	59·2	74·6
Oswego, N. Y.	304·2	34·7	62·2
Pensacola, Fla.	29·8	20·0	36·2
Philadelphia, Pa.	52·4	99·0	95·0
Phœnix, Ariz.	1,068·0 B.	3·6	18·7
Pike's Peak, Col.	14,134·2	5·1	1·0
Pilot Point, Tex.	800·0	17·2	35·6
Pioche, Nev.	6,220·0 B.	5·0	22·6
Pittsburg, Pa.	762·2	87·7	85·6
Port Eads, La.	7·1	7·0	2·0
Port Huron, Mich.	632·9	80·0	63·0
Portland, Me.	45·4	27·7	76·7
Portland, Ore.	67·0	44·9	59·8
Portsmouth, N. C.	No bar.	8·1	29·4
Prescott, Ariz.	5,339·0 B.	10·1	4·6
Punta Rassa, Fla.	13·1	13·8	35·4

Elevation of Signal Barometers above Mean Sea-Level, etc.—(Continued.)

STATION.	AB'VE SEA-LEVEL	ABOVE-GROUND.	
	Barometer.	Thermometer.	Rain-gauge.
	Feet.	Feet.	Feet.
Red Bluff, Cal	323·9	20·8	31·9
Rio Grande City, Tex	....	4·9	·0
Rochester, N. Y	588·9	100·0	96·5
Roseburg, Ore	537·0	20·0	32·8
Sacramento, Cal	69·6	37·6	54·4
St. Louis, Mo	567·8	104·4	100·0
St. Michael's, Alaska	30·0	13·0	·8
St. Paul, Minn	810·9	32·0	58·0
St. Vincent, Minn	804·0	8·8	14·0
Salt Lake City, Utah	4,348·0	52·5	74·6
San Antonio, Tex	675·7	17·2	32·8
San Diego, Cal	67·1	19·0	30·5
Sandusky, Ohio	638·6	54·0	66·1
Sandy Hook, N. J	27·9	14·9	1·0
San Francisco, Cal	60·4	48·0	75·0
Santa Fé, N. M	7,046·0	20·8	18·5
Savannah, Ga	86·9	41·0	57·7
Shreveport, La	226·8	33·3	43·8
Silver City, N. M	5,796·0 B.	4·8	1·0
Sitka, Alaska	63·0	64·4	95·6
Smithville, N. C	33·7	17·6	35·4
Socorro, N. M	4,564·8	5·2	13·4
Springfield, Ill	644·0	38·8	60·9
Springfield, Mass	120·5	54·2	63·6
Spokane Falls, Wash	1,910·0	18·5	....
Stockton, Tex	3,063·0 B.	5·8	1·5
Thatcher's Island, Mass	48·1	7·0	4·0
Toledo, Ohio	651·2	63·5	104·7
Tucson, Ariz	2,404·0	6·3	15·7
Umatilla, Ore	384·0 B.	7·3	3·0
Unalaska, Alaska	13·4	15·3	1·3
Uvalde, Tex	891·0 B.	3·6	On ground.
Vicksburg, Miss	242·6	32·5	52·6
Virginia City, Mont	5,810·0 B.	24·2	29·0
Visalia, Cal	348·1	22·4	44·5
Washington, D. C	104·6	44·1	50·8
Wickenburg, Ariz	1,400·0 B.	4·5	2·1
Wilmington, N. C	52·0	28·0	44·8
Winnemucca, Nev	4,327·3	7·0	5·0
Wood's Holl, Mass	35·0	6·1	34·0
Yankton, Dak	1,228·4	19·8	28·4
Yuma, Ariz	140·8	5·1	26·0

VI.

ETIOLOGY.

"A GLANCE at the sketch of the distribution of phthisis over the globe does not permit us to doubt that *circumstances of climate* are on the whole of merely subordinate importance for the lines of that distribution; that the disease occurs, *cæteris paribus*, in all geographical zones with uniform frequency; that equatorial and subtropical regions are visited with consumption not less than countries with a temperate or an arctic climate; that the differences which come out on comparing the amount of the malady in several parts of a given zone are of the same kind as in all other zones; that in many regions the number of cases has gone up considerably without any corresponding changes in the climate, but under circumstances of another kind" (i. e., crowded population, in-door occupation, etc.), "and, accordingly, that the notion, which is still prevalent, of a preponderance of the disease in cold or temperate latitudes, is just as erroneous as the other opinion which has lately come to the front, that consumption is found with especial frequency in those very regions that have a warm or hot climate." *

* Hirsch's "Hand-book of Geographical and Historical Pathology," vol. iii.

Cold does not produce consumption, and warmth gives no protection against it. The mean level of temperature, therefore, has no significance for the frequency or rarity of phthisis in any locality. "But it exercises a very decided influence on the course of the disease, for, according to nearly all the authorities in tropical countries, including India, Cochin-China, China, the Pacific Islands, Panama, Guiana, Brazil, and Peru, it runs usually a much more rapid and pernicious course in these countries than in higher latitudes, the removal of the patient from such a climate as speedily as may be being in fact the only sure protection against a rapidly fatal issue." * Hirsch not only shows that the absolute height of the temperature has no determining influence upon the frequency of phthisis in a locality, but also that severe and sudden changes of temperature from day to day have just as little on their own account. Evidence of this is found on a large scale in a number of places on the eastern slope of the Rocky Mountains and from the elevated prairies in the western United States. We must, however, take into account the important fact that these regions have an absolutely *dry climate*, for the circumstances are entirely different in the case of a *damp climate* subject to great variations of temperature. It is an established fact that, where we have a moist climate, along with a great range of temperature, catarrhal affections of the bronchial mucous membrane are very frequent. Hirsch states that this climatic influence is in

* Hirsch.

all probability a real predisposing factor in the development of phthisis. Dryness and variability of temperature—i. e., elevation—are, according to Dennison,* "the most important elements in the climatic treatment of phthisis."

According to Hirsch, "There are few countries of the world so characterized by uniformity of temperature and comparative dryness of the air as the inland districts of Lower Egypt and the valley of the Nile in central and Upper Egypt, regions in which phthisis, according to all observers, is very uncommon. On the other hand, localities on the coast, such as Alexandria, Damietta, and Port Said, with a moist climate and a great range of temperature, are much subject to the disease."

Hunter says, "The localities specially distinguished by dryness of climate and uniform temperature, be they on the plain or among the hills, are least affected by phthisis."

Humidity *per se* is not a factor for the production of consumption. The Hebrides, Shetlands, Iceland, and the Faroe Islands, are continually bathed in moist sea-air, and they all enjoy a remarkable immunity from consumption.

Lindsay † says, "The remarkable rarity of consumption in the navy (British) as compared with the army is a fact quite incompatible with the prevalent notion that breathing moist air predisposes to the malady."

---

* "New York Medical Journal," September 13 and 20, 1884.

† "Climatic Treatment of Consumption," by Dr. J. A. Lindsay.

Bowditch * seems to have been the first to institute inquiries in reference to the proportion of phthisis to the degree of saturation of the soil: "They were instigated by the discovery that the residents in certain localities or quarters, or even in certain houses, suffered from phthisis in a peculiar degree, while places around, even those in most immediate proximity, were less affected by the disease or unaffected by it altogether. These inquiries, carried out as they were at various places, served to show that the number of cases was in direct ratio to the degree of dampness of the soil, that the disease was found to be least prevalent upon dry soil, the drainage of the ground having been followed by a decrease in the number of cases, and that it is a matter of absolute indifference whether the saturation of the ground proceeds from one source or another." The conclusions of Bowditch have been corroborated by Milroy, Buchanan, Pepper, and other prominent observers.

"Elliott † believed that the cause of the enormous prevalence of consumption in New Orleans is to be looked for in the influence exercised by a soil saturated with moisture ('the water-level of the soil is coincident with the surface of the soil')." In this connection it is interesting to note the influence of drainage in lowering the mortality from consumption in the District of Columbia. The tenth census (1880) of the United States shows a mortality from phthisis for the city of Wash-

* "Transactions of the Massachusetts Medical Society," 1862, vol. vi, page 2.

† Hirsch, *loc. cit.*

ington of 41 for every 10,000 inhabitants, and for the rural (unsewered) portion of the District of Columbia is shown a mortality of 57 for every 10,000 inhabitants. The bad effects of aggregation are here more than balanced by the benefits of drainage.

"In a like sense Hermann * accounts for the somewhat heavy mortality from phthisis in St. Petersburg. Reeves thinks that the remarkable increase of consumption in Melbourne can be explained by the extensive irrigation lately introduced in the vicinity of the town. Devertil holds that the high death-rate from phthisis in Södermanland, particularly in the basin of Lake Mälar, while it is in part due to the great poverty of the district, is partly to be attributed to the wetness of the whole valley, the subsoil being for the most part glacial clay, which induces a high level of the ground-water, slow subsidence of the rain-water, fogs, and dampness of dwellings; and he certainly finds support for that opinion in the fact that in all the provinces south of the Dal-elf, in which the deaths from consumption reach the average or exceed it, the subsoil is glacial clay with stratified marl or stratified clay, whereas the parts of the country that have a small phthisis mortality rest upon rock or pervious gravel.

"Noteworthy as these facts must always be, it can not be overlooked, at the same time, that the rule deduced from them admits of considerable exceptions. Buchanan himself had to admit exceptional cases—for

* Hirsch, *loc. cit.*

example, that of Ashby-de-la-Zouch, where the mortality from phthisis rose 19 per cent after the ground was drained. It was subsequently pointed out by Pearse that, in several districts of Devonshire, where the rainfall was very considerable, the deaths from consumption were comparatively few, and that, if it were contended that these districts were on a pervious soil, and that other districts on wet clay showed a far more unfavorable death-rate, it was still a very remarkable thing that the mortality from phthisis was less at Wisbeach, in the fen district, than at Axminster, on the red sandstone, in which part of Devonshire, as well as in others equally fortunate in their geological foundations, but with lace-making as the industry, consumption was more common among the women (who followed that occupation amid bad ventilation) than among the men; while under hygienic circumstances that are as good as these are bad, as, for instance, on Dartmoor, the mortality from phthisis is very much less. Further, it is pointed out by Droeze, for Holland, that phthisis is far from taking a prominent place in the mortality, despite the extreme wetness of the soil everywhere; that no definite relationship can be made out with the more or less considerable wetness of the ground on comparing the mortality in the various Dutch provinces; that, in fact, the more elevated provinces with diluvial soil suffer more than the deep depressions with an alluvial soil, such as Zealand, which has the smallest phthisical death-rate (1·87 per 1,000 inhabitants). According to Reck, the mortality from consumption in Brunswick has not been greater in the wet parts of the town than in the quarters on a dry soil. In Dantzic, where a system of main drainage was carried out fully in 1871, the death-rate from phthisis, which had been (according to Lievin) 2·12 per 1,000 in the eight years preceding (1863–'70), rose in the nine years following (1871–'79) to 2·48 per 1,000. That the ground was drained by the system of sewers is beyond all question, and yet the amount of phthisis not only did not fall thereupon, but went up 17 per cent; so that Lievin concludes:

"'According to Dantzic experience, any connection between the prevalence of consumption and main drainage, as affecting the subsoil water, is certainly not made out.' At Berlin, in like manner, no notable effect on the prevalence of phthisis can be traced to drainage of the ground following the canalization of the city. Previous to 1875, in which year the canalization began, the mortality from phthisis per 1,000 inhabitants was 3·6, 3·7, 4·3, 3·8, 3·4, 3·3, and 3·5 in the successive years from 1869 to 1875. Subsequent to the latter year the annual averages were 3·3, 3·3, 3·5, 3·5, 3·5, 3·3, 3·3, 3·5, and 3·5, successively from 1876 to 1884. It is Buchanan's opinion that the exceptions to the rule worked out by him are not to be explained by errors in the observations, but that they indicate the presence of other influences in the subsoil, which have hitherto escaped detection.

"The most striking fact in connection with the geographical distribution of consumption is the rarity of the disease at great altitudes. The observations published by Archibald Smith and Tschudi as to the extreme rarity of phthisis on the high plateaus of the Andes in Peru, and as to the good effects upon the phthisical of a residence thereon, were the first statements to direct general attention to the comparative immunity from consumption of regions at a great elevation. Further inquiries in the same direction have confirmed the general fact; but they have in part also given color to an opposite conclusion, so that the question may be said to be still a *lis sub judice* for those who would decide it absolutely and without regard to accessory circumstances.

"It is not to be denied that phthisis does occur at the highest inhabited spots on the globe, and that it is rare in many places situated on low plains. None the less is it an incontestable fact that consumption is, *cæteris paribus*, much less frequently met with at high-lying places than in those at a lower elevation or on the sea-level. Not only so, but the number of cases stands in some kind of definite proportion to the degree of elevation, while the exceptions to the general rule find satisfactory explanation in other etiological factors coming into play at the same time.

"The rarity of phthisis at high elevations comes out on a great scale in the returns of sickness from that most extensive of the earth's mountain-chains which runs along the whole Pacific coast of the Western Hemisphere. For the Rocky Mountains of North America we have evidence of the fact from a number of places in the Territories situated toward the southern end of the range, such as New Mexico, Arizona, Colorado, and also Utah. In like manner all the authorities speak of the rarity of the disease on the plateaus and mountain-slopes of Mexico, Guatemala, Salvador, Costa Rica, and Panama (for example, on the Cordilleras of Veragua and Chiriqui). From Bogota, in New Granada, Holten writes that he did not see one consumptive person in the hospitals of the town during a prolonged residence there.

"For the Peruvian Andes we have the statements of Smith and Tschudi, already mentioned. During a year's stay on the Cerro Pasco the former saw only one case of consumption, and that was in a woman who had come from Europe. In those parts of the Argentine Republic that are within the limits of the Andes, the influence of high elevations upon the rarity of phthisis is observable as far down as Salta. It is still more obvious in the elevated valleys on the western side as well as on the Bolivian plateau, at Chuquisaca, Cochabamba, Potosi, and other places. In the mountainous parts of Guiana, also, consumption is almost unknown.

"In the Eastern Hemisphere this immunity from phthisis comes out most decidedly on the plateau of Armenia, where the disease is found almost solely among those who have come from less elevated places; also, on the table-land of Persia, where it is extremely rare, and among the natives of the country almost unknown; on the northern and southern slopes of the Himalayas, at the elevated points of the Western Ghâts, on the Nilghiri Hills, on Mount Abu (4,000 feet) in the Arawalli range, and in Nearer India; on the plateaus of Abyssinia, and those of Southern Africa.

"In Europe a certain rarity of phthisis begins to be noticeable even at comparatively small elevations, as in the Iser range and on the northern spurs of the Carpathians in upper Silesia, on the elevated plain of Thuringia, in the upper Harz, and in the Spessart. Writing of upper Silesia, Virchow says: "Although I have seen an exceptionally large number of sick persons of the poorer class, both in town and country, at their homes and in hospital, yet there has not come under my notice a single case of phthisis, and the statements of the medical men bear out the notion that the disease is rare." In the upper Harz consumption is so unusual that Brockmann, during a practice of many years and extending to 80,000 sick persons, found only 23 phthisical patients, of whom only 14 had been born in the upper Harz. In the lower valleys the malady is more common, but the high plateau is almost exempt. In the Spessart, according to Virchow, phthisis is at all events rare. In the larger villages he met with only an occasional case, while the registers of deaths rarely contained the entry of consumption or decline. I shall add here the interesting note by Gross, that consumption is almost unknown in Briançon (Hautes-Alpes), the most elevated town in Europe (1,306 metres or 4,285 feet), although the place is a small fortress, with a good deal of filth and a number of industries.

"Statistical inquiries, such as have been carried out in Saxony, Baden, and Switzerland, on the amount of phthisis at elevated places as compared with low-lying places close at hand (due regard being had to any differences in the mode of life), have confirmed that law of immunity of the more elevated places from phthisis which had been deduced from the study of the higher elevations by themselves. The following is Merbach's table for Saxony, based on a period of three years, from 1873 to 1875, and including only those towns with upward of 5,000 inhabitants and only the ages between fourteen and sixty:

Altitude in metres (3¼ feet).	Deaths from phthisis per 1,000 within the limit of age.
100 to 200	4·9
200 to 300	3·3
300 to 400	3·2
400 to 500	3·5
550 to 650	3·3

"Merbach concludes as follows: 'There is certainly nothing shown here of any marked influence due to the elevation of the various localities or of such an influence as would cause the number of deaths from phthisis to decrease *pari passu* with the increase in elevation. A result of that sort was indeed not to be looked for, inasmuch as the several towns are subject to other influences and some of them noxious ones, such as the occupation of the inhabitants, the density of the population, and the like, which are capable of neutralizing the good effect of an elevated location. At the same time, even in the instances before us, the good effects (otherwise sufficiently proved) of a high situation upon the prevalence of consumption can hardly be overlooked whenever we begin to compare the villages in the lowest situation with those in the highest. . . . The contrast comes out with special clearness when the averages calculated for towns situated at one and the same level are compared together.'

"Corval has worked out this relationship from the Baden bills of mortality over a period of four years (1869–'72), including in his total, as he was bound to do, all those cases where the cause of death was given as 'tuberculosis,' 'chronic pneumonia,' or 'phthisis.' He distinguishes six groups of localities according to elevation:

Table of Mortality from Phthisis in Baden, according to Elevation.

	Elevation in feet.	Number of towns or villages.	Population (average of 4 yrs.).	Deaths from phthisis per 1,000.
1	330–1,000	750	933,773	3·36
2	1,000–1,500	337	224,210	2·75
3	1,500–2,000	160	81,066	2·60
4	2,000–2,500	190	104,289	2·75
5	2,500–3,000	97	59,155	2·33
6	Above 3,000	47	20,367	2·17

"In order to ascertain what effect is produced upon the death-rate from phthisis by density of population, industrial pursuits, and other things peculiar to towns, we may make a calculation of the mortality according to the size of every town or village in Baden, using Corval's figures. We shall find, accordingly, that it is 3·12 per 1,000 inhabitants for the whole of Baden, 3 for villages of 3,000 and under, 3·49 for towns from 3,000 to 10,000, and 4·56 for towns with more than 10,000 inhabitants.

"If, now, we arrange the places that are respectively over and under 3,000 population in two columns, classifying them in the six groups according to elevation, we shall get the following table of the death-rate from phthisis:

Altitude groups.	Under 3,000 population.	Over 3,000 population.
1........................	3·11	4·05
2........................	2·73	3·08
3........................	2·49	4·99
4........................	2·71	4·72
5	2·29	3·06
6........................	2·17	....

"In the series with less than 3,000 inhabitants the favorable influence of increasing elevation is quite obvious; but in the second column of death-rates it will be seen that the benefit is in some circumstances more or less neutralized by detrimental factors belonging to the social and industrial life of the larger centers or the towns. Still, from the facts such as they are we may adopt Corval's conclusion that 'cases of phthisis decrease with increasing elevation, or, in other words, in mere increase of altitude we may discover one of the most important factors in checking the development of consumption.'

"Müller's inquiries into the effect of elevation upon the prevalence of phthisis in Switzerland have led him to the same conclusion; although the results, as he is careful to explain, can be said to be only approximately correct, for the reason that the data at his service were not free from a good many omissions and errors. He distinguishes three groups of places: (1) Those in which 43 to 63 per cent of the inhabitants follow some industrial occupation (cantons of Outer Appenzell, Glarus, Neuchatel, town and country divisions of Basel, and Geneva); (2) where the industrial part of the population reaches from 31 to 43 per cent (cantons of Zürich, St. Gall, Thurgau, Zug, Inner Appenzell, Aargau, Schaffhausen, Solothurn, Bern, Ticino); and (3) the agricultural cantons where the industrial population is only 13 to 26 per cent (Lucerne, Schwyz, Unterwalden, Vaud, Freiburg, Grisons, Uri, Vallais). Grouping the places in each of these divisions according to their elevation, within a limit of 200 to 1,800 metres (650 to 6,000 feet), we get the following table of death-rates:

Table of Death-Rates from Phthisis in Swiss Towns and Villages.

Elevation (in metres).	Industrial cantons.	Mixed cantons.	Agricultural cantons.	Average.
200–500	2·7	1·85	1·4	2·15
500–700	3·0	1·55	1·2	1·9
700–900	1·35	1·7	0·7	1·0
900–1,100	1·5	1·9	1·9	1·2
1,100–1,300	2·3	2·3	0·7	1·9
1,300–1,500	....	1·4	0·6	0·8
1,500–1,800	....	1·3	0·7	1·1
Average...........	2·55	1·7	1·1	1·86

"Müller concludes from these facts that in Switzerland consumption can be shown to decrease as we ascend; that the malady does occur, although rarely, at the highest inhabited spots; that the lowest localities have, on the average, about twice as many consumptives as the highest, and very much more than that if cases where the phthisis had been acquired elsewhere be subtracted; that the decrease of phthisis with ascending elevation is, however, neither constant nor proportionate; and that the irregularities and fluctuations which are noticeable are owing mostly to the position in the social scale, inasmuch as the industrial groups of places show the irregularities most, and the mixed groups on the whole a regular decrease with height, while the agricultural groups touch their lowest death-rate at a comparatively small elevation.

"What the minimum of elevation is that a locality must have before it feels the good effects of altitude on the prevalence of consumption is a question that can not be answered with certainty from the facts before us. Gustaldi puts it at 600 to 1,000 metres (2,000 to 3,300 feet). It seems to me, however, that a notable decrease in the disease can be shown to occur at as small an elevation as 400 or 500 metres (1,500 feet), provided other circumstances are favorable.

"Opinions differ as to the nature of the influence of altitude. Some trace it to the air being free from decomposition-products—dust and the like—others to the dryness of the air and of the soil.

"The only explanation that I can offer, and one to which I shall hold until something more satisfactory presents itself, is that people who have been born and brought up at great elevations have been always under the necessity of making frequent (or perhaps deep) inspirations, as a consequence of breathing a rarefied atmosphere; that they are continually practicing a kind of pulmonary gymnastics, from which there proceeds a vigorous development of the breathing-organs, and a greater power of resistance on their part to noxious influences from without. 'After looking at the bustle of traffic in towns like Bogota, Micuipampa, Potosi, and such like, at elevations of 8,000 to 12,000 feet,' says Boussaingault; 'after witnessing the strength and marvelous skill of the *toreadors* in the bull-fights at Quito, 9,000 feet above the sea-level; after seeing young and delicate girls dancing a whole night at places almost as high as Mont Blanc, on which the celebrated Saussure had hardly strength enough to use his instruments of observation, and his hardy guides fell down in a swoon as they proceeded to dig a hole in the snow; when we remember, finally, that a famous battle, that of Pichincha, was won almost in the altitude of Monte Rosa—I think that you will agree with me that man can become adapted to breathing the rarefied air of the very highest mountains.' I will readily grant that many of the accounts of embarrassed breathing experienced by natives of the plains on ascending very high mountains are exaggerated; and I must confess that, in my own case, after resting for a short time at elevations of 10,000 feet and upward, I was conscious of no considerable want of breath, or did not become aware, at least, of any need for quicker or deeper inspirations. At the same time, it is not to be denied that the atmosphere at elevations of 10,000 feet, especially in a warm climate, is rarefied to the extent of more than one third of its volume at the sea-level. The quantity of oxygen contained in it is, therefore, considerably diminished, and a man must take in a larger quantity of air in a given time, or must inspire oftener than on the plains, so as to cover his requirements for oxygen. To that assumption no well-grounded objection can be raised, whether from the side of physics or of physiology, and there is equally little reason why we should not assume that those who have been born and have lived all their lives under such circumstances will have had their breathing-organs powerfully developed. I do not hesitate, therefore, to discover the reason of the immunity from phthisis enjoyed by the residents of elevated places, in the influence which a continuous residence in a rarefied atmosphere exercises over them."

Bell * says, "The special effects of rarefaction of air upon the animal economy has been illustrated and most satisfactorily studied during and since the construction of the Oroya Railroad, by Mr. Henry Meiggs, the only one in the world reaching the altitude of 15,640 feet." He gives the following particulars regarding the effects of rarefied air from his own experience on this road, and from the written and oral statements of Dr. G. A. Ward, who has been employed professionally on the road since its beginning:

"The labor employed in building this road was principally the native Peruvians of the mountains, who are a short, thick-set race, called *serranos*, and have immense lung-capacity. Mr. S. W. North, civil engineer, made some measurements of these *serranos* at Yauliyacu, an altitude of 16,000 feet, as follows:

---

* "Climatology," etc.

AGE.	Chest measurement.	HEIGHT.		
		Proper height in inches, of twice the chest measurement, European standard.	Actual height in feet and inches, and in inches alone.	Difference in inches.
	Inches.	Inches.	Ft. in. in.	
14 years...........	36	72	4 10 = 58	14
24 years...........	36	72	5 6½ = 66½	5½
21 years...........	35	70	5 4 = 64	6
16 years...........	34½	69	5 0 = 60	9
30 years...........	34½	69	5 4½ = 64½	4½

Average difference in height between European and *serrano*, 7⅛ inches.

"This enormous increase in size of the chest is owing to the rarefaction of the air in which these natives live, enabling them to undergo an active and even laborious existence at these great altitudes. American engineers employed in building the road increased their lung-capacity during their labors. One of these, Mr. John Malloy, informed me that the measurement of his chest had been increased four inches in two years by exposure to rarefied air in these Andes.

"This peculiarity of adaptation to the demands of Nature enables these people to overcome the pains and inconveniences which are experienced by persons who ascend the Andes for the first time toward their summits, and which are known under the names of *soroche*, *veta*, *puna*, etc. These symptoms indicate a diminished supply of oxygen, but more particularly a diminished pressure of air on the surface of the body and on the interior of the lungs.

"The pressure at the sea-level, constantly diminishing as you ascend, is found to be reduced to about one half at an elevation of 16,000 feet, or the summit tunnel of the Oroya Railroad. This withdrawal of pressure often occasions the most severe symptoms of vertigo, headache, nausea, and vomiting, all more or less alarming, and attended with profound prostration. The whole are attended with increased respiration and rapid action of the heart. Dr. Ward says some are affected with fearful nausea and vomiting, comparing it to the worst form of sea-sickness. Others suffer from severe frontal headache, palpitation of the heart, etc. From the violence of the heart's action it really seems at times as if it would burst the walls of the chest. Occasionally severe hæmorrhage occurs from all the avenues of the body.

"The respirations are increased from three to five times a minute. Dr. Ward says he has counted 43 respirations and 148 pulse-beats in a minute at an elevation of only 9,000 feet, and that the pulse is *always* increased in frequency but not in volume. A person who at the sea-level has a pulse of 75 per minute would find it increased about ten beats at an altitude of 10,000 feet, and would experience ten additional beats for each 1,000 feet of added altitude. The rule is that no one passes for the first time an altitude of 16,000 feet whose pulse does not mount to from 130 to 150 beats in a minute.

"These increased numbers of pulsations are absolutely necessary to avert a fatal result. The attending increased respiratory action is not accompanied with increase of secretions, but an increased amount of air of inferior quality, from actual reduction of the amount of oxygen, fails to aërate or properly preserve the fluidity of the blood. . . . Mountain-air of an altitude of 2,500 feet and upward, with rare exceptions, possesses the one chief attribute of salubrity common to sea-air—freedom from organic impurities. Pasteur, Tyndall, and others have shown that the air of great altitudes is entirely free from organic impurities.

"Miguel, as recently quoted by Weber, gives the following interesting table of the number of bacteria found in ten cubic metres of air taken as nearly as possible at the same time in July, 1883:

1. At an elevation of from 2,000 to 4,000 metres........	0
2. On the lake of Thun (560 metres)........................	8
3. Near the Hôtel Bellevue, Thun (560 metres)..........	25
4. In a room of the Hôtel Bellevue.......................	600
5. In the Park of Montsouris, near Paris...............	7,600
6. In Paris itself (Rue de Rivoli)........................	55,000

"This table is doubtless equally indicative of the difference in the amount of floating organic matter in the air at different altitudes."

* "Differences in the *social, hygienic, commercial, and industrial conditions* of various parts of the world have a real import for the more or less frequent occurrence of phthisis therein. This is shown in the first instance by the distribution of the malady respectively in *town* and *country*, in large populous towns and in those that are small or thinly populated, and among a stationary or a nomadic population. As a general rule, phthisis is commoner in towns than in the open country, and rarer in small towns than in large, or, in other words, it

* Hirsch.

is found in greater diffusion the more crowded the population.

"In England, according to the statistics of mortality from 1859 to 1869, the lowest death-rates from phthisis (1·8 to 2·2 per 1,000) are found in the counties most given up to agriculture and pasture, and with few large towns, such as Rutland, Worcester, Wilts, Dorset, Somerset, Herts, Bucks, Hereford, Gloucester, Shropshire, and Lincoln; while the highest death-rates (2·7 to 3·3 per 1,000) occur in the counties with many places of manufacture and trade, the centers of industry and commerce, such as Notts, Derby, West and East Ridings, Durham, Northumberland, Cheshire, Lancashire, and the metropolis. For Denmark the following ratios of death from phthisis have been calculated by Lehmann from the statistics of 1876–'83: In Copenhagen, 3 per 1,000; in the five largest provincial towns, mostly engaged in trade, shipping, and manufacture, and with populations from 12,700 to 25,000, 2·63 per 1,000; in twenty-four towns, with from 3,100 to 9,000 inhabitants, agricultural in the first place but also concerned in trade, manufacture, and handicrafts, 2·27 per 1,000; in the thirty-five smallest towns, with from 700 to 2,900 inhabitants, mostly agricultural, but also occupied with fishing and shipping, 2·12 per 1,000. In Holland, according to Fokker, the death-rate from phthisis in the towns is to that in the open country as 21 to 16. Summing up his statistical inquiries on the mortality for the years 1866–'75, Droeze says, "In nearly every group of places in the most diverse provinces of Holland the death-rate from phthisis was greater in the towns than in the country parishes.

"The following table, which has been compiled from Schlockow's figures, showing the mortality from phthisis in town and country in the several administrative divisions of Prussia, is of value in this connection:

Mortality from Phthisis in Town and Country, per 1,000 Inhabitants (Prussia).

	Town.	Country.
Königsberg	2·49	1·45
Gumbinnen	2·77	1·84
Dantsic	2·39	1·41
Marienwerder	2·54	1·35
Potsdam	2·88	2·33
Frankfort	3·08	2·25
Stettin	2·90	2·08
Köslin	2·58	1·60
Stralsund	3·21	2·12
Posen	2·96	2·04
Bromberg	3·13	1·85
Breslau	3·73	2·75
Liegnitz	2·98	2·35
Oppeln	3·90	2·45
Magdeburg	2·98	2·65
Merseburg	2·63	2·16
Erfurt	2·69	2·70
Schleswig	3·31	3·18
Hanover	3·38	4·44
Hildesheim	2·66	3·21
Lüneburg	3·85	3·39
Stade	3·18	4·20
Osnabrück	4·87	5·22
Aurich	3·81	3·79
Münster	6·50	4·70
Minden	4·73	4·90
Arnsberg	5·46	4·51
Cassel	3·48	3·03
Wiesbaden	3·82	4·08
Coblentz	4·26	4·35
Düsseldorf	5·22	5·29
Cologne	4·76	5·34
Treves	3·53	3·56
Aix	3·64	4·59
Sigmaringen	3·11	3·63

"In Baden, from 1852–'71, the death-rates from phthisis group themselves as follows:

Villages or towns.	Population.	Deaths from phthisis per 1,000.
95....................	100– 500	2·3
92....................	500– 1,000	2·6
53....................	1,000– 2,000	3·0
17....................	2,000– 4,000	3·4
3....................	4,000– 8,000	5·5
3....................	17,000–31,000	3·9

"In the Bavarian administrative section of middle Franconia, Majer estimates the proportion of deaths from phthisis between town and country at 100 to 61. In the arrondissement of Dax (département Landes) consumption is hardly known among the rural population, but it is far from rare in the towns, and the same fact is recorded of the arrondissement of Nerac (département Lot-Garonne). In the course of inquiries upon the antagonism between phthisis and malarial diseases in Ferrara and vicinity, Bosi and Gambari came to the conclusion that phthisis was rare among the rural population at those places where malarial fevers were endemic, that it was more common in the villages with a rather crowded population, and that it was very prevalent in the large towns. They found the same proportions also in those parts of the country where malarial fever was merely sporadic; in other words, the density of the population was, *cœteris paribus*, decisive for consumption. It is very significant for the question before us, as already indicated more than once, that nomad peoples enjoy an almost complete freedom from phthisis; this holds good for the Kirghiz hordes of the Russian steppes, for the Arab tribes in Kabylia and other parts of Northern Africa, and for those of nearer Asia. The disease is met with oftenest, says Pruner, 'among those Bedouin families who have exchanged the tent for the dwelling built of stone.'"

Another interesting proof of the influence exercised upon the amount of phthisis by the social factors is afforded us in the disease increasing as a result of extensive *immigration* and the consequent *founding of new towns, or enlargement and crowding of old ones.*

Writing in the last century, Rush * says: "Phthisis is scarcely known by those citizens of the United States who live in the first stage of civilized life, and who have lately obtained the title of the first settlers; it is less common in country places than in cities, and increases in both with intemperance and sedentary modes of life." Since that was written, the disease has increased considerably, not only in the Eastern States but also in the Western, along with the founding of cities and the rise of traffic and industry. Writing in 1828, of the western counties of Pennsylvania, Callaghan says that "phthisis is increasing among the sedentary population of our towns with fearful strides"; and for a more recent period there are accounts from that State, as well as from California, of the disease increasing hand in hand with progressive immigration and additions to the

* "Medical Inquiries and Observations," Philadelphia, 1789, p. 159. Hirsch, *loc. cit.*

population of the towns. Still more recently Davis has written of the Western States as follows: "Close buildings and increased aggregation of population are increasing the prevalence of pulmonary tuberculosis in our country at an alarming rate. . . . In still earlier days, dating back to the early settlements of this country, New England and the Northeastern States were as free from consumption as are now the much-vaunted far-Western States and Territories. It was immediately consequent upon the change from an agricultural to a manufacturing population that the rapid increase in the death-rate from consumption is apparent in these States. Fifteen or twenty years ago Indiana, Illinois, and the lake region were the favorite resorts for consumptive patients. . . . Now we have a constantly increasing proportion of cases originating in this same region, not evidently from any change that has taken place in the climatic conditions, but, as before stated, from the change in the occupation and hygienic surroundings of the people."

Flick * discovers in the Americanizing influences to which our inhabitants of foreign parentage, particularly the Irish, are usually subjected, a potent factor in the etiology of consumption. Although his views on the subject, which I give in his own words, may be incorrect, they nevertheless possess the merit of being novel:

---

* "The Hygiene of Phthisis," a paper read before the Philadelphia County Medical Society, January 11, 1888, by Lawrence F. Flick, M. D., *loc. cit.* The "Journal of the American Medical Association," February 4, 1888.

"According to the United States census reports for 1880, . . . among people of Irish parentage, 309,507 males and 375,636 females die of consumption to every million deaths; and among people of German parentage the victims of the disease number 249,498 males and 254,958 females to every million deaths. It will be seen that the largest percentage of deaths from the disease is among Irish immigrants and their children. This is usually ascribed to the change in climate. Ireland has a much damper climate than America, and therefore one better suited to the development of phthisis. The real cause for the larger mortality from consumption among foreigners, and especially among the Irish, is the change in diet. At home they have been accustomed to a plain, healthy diet, and when they come to this country they at once take to the varied heavy diet of Americans. Where they have eaten little meat at home, they eat it in profusion here; where they have drunk good milk and eaten vegetables at home, they drink teas and coffees and eat spiced foods here. They soon become thorough Americans in their stomachs, and even outdo the natives. The consequences are indigestion, malnutrition, tuberculosis, etc. The German, though frequently pursuing a similar course, is often spared by his characteristic thrift and economy. He partakes more sparingly of the good things that come in his way, because of his anxiety to prepare for a rainy day. His fondness for beer, a beverage that he manages to secure wherever he goes, may likewise have some influence in shielding him against phthisis."

Hirsch believes that "the exceedingly common occurrence of phthisis among the new arrivals themselves is in part the explanation of the progressive increase of phthisis in the cities of the United States, that goes on hand in hand with immigration on a large scale; but a not less considerable part of it is due to an increase of the malady among the settled population of the towns. It must be connected, therefore, with influences of a general kind proceeding from a change in the mode of living within the current century, and particularly within the more recent years of it, with *social errors* such as mostly obtain in large towns and in the centers of industrial traffic —errors from which no city or manufacturing town can escape. Among those detrimental things in the social life, we have, according to nearly unanimous opinion, to assign the first place, along with insufficient or bad food, to the *bad domestic hygiene*, to the influence of continuous residence in crowded living-rooms and work-rooms, tainted with organic and inorganic exuviæ, ill ventilated, and damp. In fact, it would be hard to find any factor in the production of phthisis which can claim more importance than that."

"The effect of sedentary habits," says Clark, "in all classes and conditions of society is, in my opinion, most pernicious, and there is perhaps no cause, not even excepting hereditary predisposition, which exerts such a decided influence in the production of consumption as the privation of fresh air and exercise; indeed, the result of my inquiries leads to the conviction that sedentary habits are among the most powerful causes of tuberculous diseases, and that they operate in the higher classes as the principal cause of its greater frequency among females." "All modes of life," says Andrew, referring to the origin of phthisis, "all occupations which are carried on indoors, contrast unfavorably with out-door pursuits. The naked savage, whatever ills he may have to bear, rarely finds phthisis among them; but with every addition to his clothing, and to the comfort of his tree or cave, his proneness to it increases." Flint sums up his many years' experience of hospital and private practice as follows:* "The general conclusion is, that occupation has an agency in the etiology of pulmonary tuberculosis, in so far as it is sedentary and involves confinement within-doors. If it be said that this conclusion is in accordance with what is already known, I answer that the correctness of the conclusion is thereby made the more certain."

Lindsay † says: "There is now a vast mass of evidence which conclusively proves that consumption is comparatively rare among those who follow an out-door life under normal and healthy conditions; that it is comparatively common among those who live habitually in-doors; and that it attains its maximum incidence among those whose occupation involves prolonged confinement in a vitiated atmosphere. Thus, of 1,000 fishermen who die, only 108 succumb to consumption, while among grocers the mortality rises to 167, among drapers it mounts to 301, and among printers attains the truly appalling total of 461.

* "New York Medical Record," January, 1873, p. 49.
† "Climatic Treatment of Consumption."

Hence, it would appear that, if a man chooses the life of a fisherman, it is about ten to one that he will not contract consumption; while, if he becomes a printer, it is almost an even chance that he will fall a victim to that disease. That the open-air life on the one hand, and the confinement in vitiated atmosphere on the other, are the essential factors, becomes evident when further statistics on the subject are interrogated. Thus, next to fishermen, the classes whose occupation involves the largest average of healthy open-air life are those engaged in the cultivation of the soil. Accordingly, we find that the mortality from consumption per 1,000 deaths is 103 among farmers, 121 among gardeners, and 122 among agricultural laborers. Farmers thus even surpass fishermen in their exemption from consumption—a fact due, no doubt, to their better position and greater immunity from hardship—while gardeners and agricultural laborers only slightly fall below them. At the other end of the scale, among the classes which suffer most from consumption, we have cutlers, with a mortality rate of 371; file-makers, whose rate is 433; and earthenware manufacturers, of whom 473 out of every 1,000 die of consumption. If we compare different in-door occupations, we find that, as the contamination of the air increases, the death-rate from consumption rises *pari passu*. Thus, the workers engaged in the hosiery manufacture spend their time in-doors it is true; but there is nothing in the nature of the work to create any special contamination of air. Hence the mortality from consumption among this class of operatives attains only the very moderate figure of 168 per 1,000, actually less than that for the community as a whole, whose rate is 220 per 1,000; but among the operatives engaged in the woolen and cotton manufacture, which involves the inhalation of particles of dust, the rate rises to 257 and 272 respectively. The apparent exceptions to the law that the mortality from consumption is in direct ratio to the contamination of the air of respiration admit, for the most part, of a ready explanation. Thus, quarrymen, although working in the open air, have a mortality rate of 308; but here comes into play the inhalation of particles of stone-dust, which we know to be one of the most powerful predisposing causes of consumption. Cab and omnibus drivers, also working in the open air, have a mortality rate from consumption of 359; but when we observe that 1,482 of this class die from all causes, as compared with 1,000 of the general population, it seems reasonable to conclude that the excessive mortality from consumption is simply a part of the general unhealthiness of the class—an unhealthiness due partly to exposure, partly to intemperance, and partly to the fact that these occupations are often adopted by men who have relinquished other trades in consequence of a breakdown in health. One notable exception to the general law is more difficult to explain, viz., the comparative exemption from consumption enjoyed by the workers in coal-mines. Their mortality rate from this cause is only about 126, or nearly as low as that of the agricultural laborers, yet they work in a confined atmosphere and among much dust. Some high authorities have explained this immunity by the assumption that coal-dust actually possesses the power of inhibiting the development of consumption. This theory is so opposed to all our knowledge of the disease that we are inclined to look elsewhere for an explanation. Two facts may perhaps give us the clew to a more acceptable hypothesis: The work in coal-mines is excessively laborious, hence it is not at all likely to be chosen by those whom hereditary tendency or acquired debility of any kind predisposes to consumption; in the second place, we know how often a sudden and marked impairment of physical vigor is the first premonition of threatening consumption—hence it is probable that many coal-miners, on becoming sensible of this diminished vitality, relinquish their laborious work, and, seeking a livelihood by some lighter occupation, fail to be tabulated as miners in the mortality returns."

Returning to Hirsch, we find the following valuable data on this subject: "Summing up Greenhow's inquiries, which were based in part on official statistics of the mortality, and in part on independent local researches into the death-rate from consumption in the manufacturing districts of England, Simon concludes as follows:

"'In proportion as the male and female populations are severally attracted to indoor branches of industry, in such proportion, other things being equal, their respective death-rates by lung-disease increased, . . . and this further conclusive proof was given as to the influence of an accused occupation, viz., that the high death-rate from lung-disease belonged, according to the occupation, to men or to women of the district, that it sometimes was nearly twice as high for the employed sex as for the unemployed sex, and that it only extended to both sexes where both were engaged in the occupation.'

"Smith has ascertained, for one thousand persons treated for consumption at the Brompton Hospital, that 70 per cent of them had been in the habit of spending their time in over-crowded, hot, and dusty places indoors. Finkelnburg's summary of his inquiries into the causes of mortality in Rhenish Prussia is as follows:

"'The victims of pulmonary phthisis are the more numerous the more generally are indoor occupations followed by the one sex or the other, especially when the materials of their work are such as to create dust—wool-carding and spinning, knife-grinding, and metal-polishing are the most pernicious. Wherever these occupations are found among the rural population as well, there also the mortality from phthisis reaches a high figure, although never so high as in towns with the same industries.'

"From the paper of Schweig on the distribution of phthisis in Baden, it appears that the mortality from it is in proportion to the density of the population, as well as to the more or less industrial character of each locality, the smallest communities being mostly engaged in agriculture, while the larger villages and large towns are occupied with industrial pursuits.

"Kolb shows, from the statistics of Bavarian hospitals for 1877 and 1878, that 'consumption in Bavaria is commonest in the highly industrial region of central Franconia, where the influence of manufactures has been at work for generations.' In Müller's inquiries into the state of health in Switzerland, he divides the country into agricultural, industrial, and mixed sections (the last being partly devoted to farming and partly to manufacture), and he finds that the phthisical death-rate of the industrial group stands to that of the agricultural in the ratio of 69·8 to 30·2; of the industrial to that of the mixed, in the ratio of 53·8 to 46·2; and of the mixed to that of the agricultural, in the ratio of 66·5 to 33·5. The general law deduced from his research is, that the mortality from phthisis in the industrial circles is, on the average, more than double that in the agricultural. The actual figures are 2·55 per 1,000 in the industrial localities, 1·7 in the mixed, and 1·1 in the agricultural, giving a proportion of 47·7 to 31·8 to 20·8. In Belgium, according to Meynne (*loc. cit.*), the highest proportion of deaths from phthisis falls to the industrial divisions of the country. From a paper by Chatin, it appears that the amount of phthisis among the factory-hands in Lyons is altogether enormous; it is greater than in any other part of France or in any other country, the mortality from it having amounted in 1866 to 33 per cent of the deaths from all causes at the Hôpital de la Croix-Rousse. This fact is confirmed by Fonteret, who says that the female part of the working-class suffers more than the male, for the reason that the women are more subject than the men to the noxious influences (sedentary indoor life in factories).

"Poulet calls attention to the fact that, in the village of Plancher-les-Mines (arrondissement Lure, départemente Haute-Sâone), where the people until about thirty years ago were occupied with agriculture almost exclusively, but have been devoted since that time mostly to industrial pursuits, phthisis now causes 12·5 per cent of the total mortality, whereas formerly it was very seldom seen. The malady is exceedingly common among the Kashmiri weavers in and around Amritsur (Punjaub), who live, as Hinder tells us, crowded together in confined, dark, and filthy rooms.

"The same circumstances serve to account for the strikingly common occurrence of phthisis in nunneries, seminaries, and such-like institutions, in evidence whereof a number of observations have been brought forward by Fourcault; also in the Oriental harems, not only among the women, but among the children also; again, among badly lodged troops, of which we have evidence from England, France, Turkey, and India; and, above all, in prisons.

"Among army-surgeons there is complete agreement that cases of phthisis are least common in soldiers when they are leading an active life in the open air, on the march, or in manœuvres and campaigns, and that the cases mount up as soon as the troops enter on their garrison-life, as, for example, in winter, and spend their time in ill-constructed, crowded, filthy, and badly ventilated barracks. Welch, who treats of this matter with reference to the British army, says that 'nearly half of army consumption is connected with vitiated barrack atmosphere,' a similar opinion having been expressed by earlier writers, such as Tulloch and Maclean, the latter including in his statement the British and native troops in India. With respect to its frequency in the French army, we find a similar reading of the facts in the papers by Champouillon, Tholozan, Viry, Lausiès, and others."

*Phthisis in Prisons.**—"Consumption prevails in prisons to a truly disastrous extent. In United States prisons from 1829 to 1845 the mortality from phthisis was 12·82 per 1,000 prisoners at Philadelphia, and at Auburn and Boston 9·89 and 10·78 respectively; in Baltimore prison it was 61 per cent of the mortality from all causes. In the French prisons, particularly those in which long terms of penal servitude are worked out, the death-rate from phthisis amounts to between 30 and 50 per cent of the mortality from all causes. In the Dutch prisons it reaches the same height; in the Danish convict-prisons it amounted in 1863–'69 to 39 per cent of all the deaths; over the whole of the prisons of the Austrian Empire in 1877–'80 it was 61·3 per cent; and in the nine large convict-prisons of Bavaria from 1868 to 1875 it was 38·2 per cent. In the penal establishments of Würtemberg, according to Cless, the yearly average of deaths from phthisis from 1850 to 1859 was 24 per 1,000; while from 1859 to 1876, in consequence of an improved diet, it fell, as we have seen, to 8 per 1,000, although it still remained two or three times greater than among the people at large. During a period of eleven years (1869–'79) the mortality in the prisons of Prussia was 42·87 per cent of the deaths from all causes, and 12·32 per 1,000 prisoners.

---

* Hirsch.

"For England we have Baly's report on the prevalence of phthisis from 1825 to 1842 among the convicts at Millbank Penitentiary, where 31 out of 205 deaths were due to cholera, and 75 of the remaining 174, or 43 per cent, due to phthisis; while of 355 prisoners discharged during the same period on account of ill health, 90 were phthisical, and of these quite three fifths, according to precedent, would have died of that disease if they had been left to complete their term. In that way we bring the annual mortality from phthisis at Millbank up to 13 per 1,000, or more than three times that of the London population at large. Pietra Santa gives the following facts for the prisons of Algiers: Of 23 natives who died in the public prison of Alger, 17 succumbed to phthisis; in the central prison of l'Harrach there were 57 deaths from phthisis in a total of 153, or 37·2 per cent. The important influence of imprisonment on the occurrence of this disease is very clearly brought out by its prevalence in those regions where phthisis is in general a rare thing, as, for example, in Lower Bengal. Webb quotes the following remarks by Green with reference to the commonness of the disease among the natives in the prison of Midnapore:

"'After a careful examination into the early history and origin of the cases of this disease as they have occurred, I have been led to the conclusion that many of the men thus affected were previously hale, and capable of earning their livelihood, and were not subject to cough before imprisonment. I find that, after they have been working a few weeks or months on the roads here, and inhabiting the jail, they have become the subjects of attacks of inflammation of the lungs, and from time to time of frequent repetition of these attacks, which have ended, in some cases, . . . in death in the acute stage, in others in a prostrate sinking state, with a gradual wasting away of the body, and all the symptoms and ultimately all the post-mortem morbid appearances of tubercular disease of the lungs.' Next to the hard labor, Green lays most stress on the bad ventilation of the cells, and on the highly defective construction of the prison in other respects

"The great frequency of consumption in convict-prisons may seem to be due to many of the prisoners bringing the disease with them; but that such is not the case follows from the well-authenticated fact that most of the deaths from phthisis among prisoners do not occur until the later years of their term of confinement. At Millbank Penitentiary signs of a pulmonary affection on admission could be made out, as Baly tells us, in only 12 prisoners among 1,502 who entered in 1842, and in only 15 among 3,249 who were received in 1844. Among the convicts of 1842 there were 510 women sentenced to transportation, who remained at Millbank not longer than three months, and of these 2 fell ill with phthisis or scrofula during that time, whereas of the remaining prisoners admitted no fewer than 47 became consumptive before the completion of their terms of two or two and a half years. It is further to be kept in mind that most of the convicts sent to Millbank had already served longer or shorter terms in smaller prisons elsewhere, and not a few of them more than one term, so that, in a certain proportion of those who were found phthisical on admission to the central prison, the seeds of the disease might have been implanted while they were undergoing sentence previously." *

Heredity.—Of real importance for the frequent occurrence of phthisis is the transmission of the disease by way of *heredity*. "That phthisis propagates itself in many families from generation to generation is so much a matter of daily experience that the severest skeptic can hardly venture to deny an hereditary element in the case, even if we be unable for the present to decide whether it consists in the transmission of a specific poison, something like that of syphilis; or, in other words, whether it be heredity in the narrower sense; or whether it does not rather depend upon a congenital disposition toward the disease, a disposition that has to be looked for naturally in the organization of the respiratory system.

"In Switzerland, according to Müller, the number of cases in which heredity was made out for certain exceeded by a little (5 or 6 per cent) those in which the malady had been acquired. Walshe found phthisis to be hereditary in 162 out of 446 families, or in one third

* In closing this subject I wish to call attention to the fact that there exists in this great country, populated as it is by sixty million people, one seventh of whom will succumb to pulmonary consumption, but one institution for the rational (pure air) treatment of the disease under consideration. An able appeal for the general establishment of such institutions in the United States was made in a paper which was read before the New York Academy of Medicine, February 18, 1888, by Paul H. [redacted] Kretzschmar M. D., but it has met with little success.

of them. In a thousand cases of consumption, Smith ascertained that the parents had been phthisical in 21·1 per cent, and the brothers or sisters in 23 per cent. Brünicke reckons the number of hereditary cases in Copenhagen at 46 per cent of the whole. Gjör was able to prove heredity in 197 out of 357 patients treated for phthisis in the hospital of Christiania, or 55·1 per cent." *

Referring to hereditary tuberculosis, Koch † says: "There are no facts to prove the view that tubercle-bacilli may be present in the immature organism, either in the intra-uterine or extra-uterine state, without leading in a relatively short time to visible changes. Now, tuberculosis is very rarely found in the fœtus and the newly-born, hence we must conclude that the infective material comes into operation only exceptionally during intra-uterine life. This view corresponds with the fact that those of the animals experimented on, particularly Guinea-pigs, which were either pregnant before or became so after inoculation, in no case gave birth to young which already showed signs of tuberculosis. The young of highly tubercular mothers were free from tuberculosis and remained healthy for months. In my opinion, hereditary tuberculosis is explained most naturally by supposing that the infective germ itself is not inherited, but rather certain peculiarities favorable to the development of germs which may later on come into contact with the

* Hirsch.

† "Investigation of Pathogenic Organisms." New Sydenham Society's Translations, 1886, p. 200.

body; in fact, it is the predisposition to tuberculosis which is inherited."

Contagious Transmission.—Hirsch, speaking of the contagious transmission of tuberculosis, says that "those who start convinced that the terms 'pulmonary consumption' and 'pulmonary tuberculosis' cover each other exactly, or that the anatomical changes proper to phthisis in the tissues of the lung depend absolutely and always upon the penetration of tubercle-bacilli into that organ, and who know or seek to know nothing else than the positive results of experiments to inoculate animals with tubercle-bacilli—such persons have no object in discussing the question of the spread of phthisis by contagious transmission; it is answered for them absolutely and unconditionally in the affirmative. But the case is different with those who proceed to solve the question from the side of actual experience (and in such matters these men have an important voice), who look at those experiences on all sides, and test their value as proofs that may be adduced in favor of the contagiousness of consumption.

"It will have been observed that there have lately been instituted in England 'collective investigations' or etiological inquiries upon a number of the more ordinary diseases, conducted in common according to a definite plan. One of these has had reference to phthisis, and has yielded the following conclusions with respect to the transmission of that disease: Of 1,078 answers to the question, 673 were simply neutral—that is to say, so many of those who returned the card had no information to give one way or another; in 105 of the answers, the question was decidedly negatived; in 39, the answer was doubtful; and in 261 cases, transmission was absolutely affirmed. Among these 261 cases, phthisis had passed from husband to wife in 119, from wife to husband in 69, from parents to children or between the children of one family in 81, to more distant relatives in 13, and to those who stood in no relationship in 8 cases. Prof. Humphry, of Cambridge, and Dr. Mahomed, of London, who edited the report of the committee on these collective investigations, do not admit any further inference from them than that, "if phthisis is a communicable disease, it is only under circumstances and conditions of extremely close personal intimacy, such as persons sharing the same bed or the same room, or shut up together in numbers in close, ill-ventilated apartments."

C. T. Williams gives the following facts relating to the Brompton Hospital for Consumption, the largest institution in the world devoted to the treatment of the phthisical:*

"The hospital has been in existence since 1846, in which year it was opened with 90 beds. In 1856 the number of beds was increased to 200, and in 1873 to 240. Three fourths of the patients suffer from phthisis in its various stages, the remainder being admitted for bronchitis, pleurisy, empyema, chronic pneumonia, and the like. Previous to 1877 the left wing was ventilated most imperfectly; since that year, however, the extraction of foul air has been well performed. The spittoons

* "British Medical Journal," September, 1882, page 618.

of the patients are changed two or three times a day; but until lately no attempt was made to disinfect them unless the odor was unpleasant.

"The out-patient department was until the winter of 1881–'82 situated in the old hospital, and was much too small for the number of patients, who averaged 200 to 300 daily, mostly phthisical. This large concourse must, on the theory of infection, have proved a considerable source of danger to the assistant physicians, to the clerk who enters their names, and to the porters who marshal them and keep order.

"The deficiency in the ventilation," says Williams, "must have led to a large accumulation in the wards of the products of respiration and also of our friends the bacilli. We consequently ought to have seen an extension of the disease to non-consumptive cases or to the nurses; but nothing of the sort occurred, only the usual results of hospitalism, i. e., erysipelas and sore throat. Among the physicians, assistant physicians, clinical clerks, nurses, and others, to the number of several hundred, who had served in the hospital (not a few of them having lived in it for a number of years continuously), phthisis had not been more common than it may be expected to be on the average among the civil population of a town; and only in three or four cases could the outbreak of it be brought in any way into connection with the individual's residence in the hospital.

"The evidence of large institutions for the treatment of consumption, such as the Brompton Hospital, directly negatives," Williams concludes, "any idea of consumption being a distinctly infective disease, like a zymotic fever." He admits that in his private practice a few cases had occurred of phthisis ensuing in those who had been in very close intimacy with consumptives; "but, when we bear in mind the far greater number of examples of consumptives living in close intimacy with healthy people, in such relationships as husband and wife, mother and daughter, or sisters sleeping together, where no spread of tubercular disease has taken place, we must admit that the negative evidence against infection greatly preponderates over that of the very few positive instances."

"During a practice of twenty-three years in an extensive district (Tynedale), Fraser had not seen a single case of consumption which told in favor of transmission from husband to wife, or *vice versa*. In twenty-six fatal cases, in which either the husband was affected or the wife, the married couple had shared the same bed and lived in the closest intercourse with each other without any transmission of the disease taking place. Over a half of these twenty-six persons had near relatives similarly affected; phthisis had already proved fatal to children of nine of these marriages, and, judging from appearances, many more were likely to suffer. Reginald Thompson has had under observation fifteen well-marked examples of wives infected by husbands out of something like 15,000 cases of phthisis, so that the proportion may be reckoned as not less than one per thousand.

"Bennet gives it as his opinion, based on twenty-five years' experience, that, if there has been any spread of phthisis at all by means of contagion, it has occurred extremely seldom, and only in quite peculiar circumstances."*

"If we inquire," according to Koch,† "how far phthisis may occasion the transference of tubercle-bacilli from diseased to healthy subjects, it is very evident that all the conditions for the distribution of the infective material in very large quantities are here present. It is necessary only to remember that on an average one seventh of mankind die of phthisis, and that most phthisical patients eject for at least some weeks, often for whole months, large quantities of sputa, containing immense numbers of spore-bearing tubercle-bacilli. Most of these countless infective germs, which are scattered everywhere, on the floor, on articles of clothing, etc., perish without finding an opportunity of settling again in a living host; but if we further bear in mind the results of Fischer and Schill's experiments, from which it is seen that tubercle-bacilli may retain their virulence for 43 days in putrefying sputum, and for 186 days in sputum dried at the ordinary temperature of the air—i. e., if we remember the immense number of tubercle-bacilli derived from phthisical patients, and, as we have just seen, their tenacity of life both in a moist and in a dry condition—a sufficient explanation is afforded of the very wide distribution of the tubercular virus.

"There can likewise be no doubt as to the manner in which the tubercular virus is carried from phthisical to

---

* Hirsch.

† Koch, "Investigation of Pathogenic Organisms," etc.

healthy subjects. By the force of the patient's cough, particles of tenacious sputum are dislodged, discharged into the air, and so scattered to some extent. Now, numerous experiments have shown that the inhalation of scattered particles of phthisical sputum causes tuberculosis with absolute certainty, not only in animals easily susceptible to the disease, but in those also which have much more power of resisting it. It is not to be supposed that man would be an exception to this rule, but, on the contrary, we may surmise that any healthy person brought into immediate contact with a phthisical patient and inhaling the fragments of fresh sputum discharged into the air, may be thereby infected. But probably infection will not often take place in this way, because the particles of sputum are not small enough to remain suspended in the air for any length of time. Dried sputum, on the contrary, is much more likely to cause infection, as, owing to the negligence with which the expectoration of phthisical patients is treated, it must evidently enter the atmosphere in considerable quantity. The sputum is not only ejected directly on to the floor, there to dry up, to be pulverized and to rise again in the form of dust, but a good deal of it dries on bed-linen, articles of clothing, and especially pocket-handkerchiefs, which even the cleanliest of patients can not help soiling with the dangerous infective material when wiping the mouth after expectoration, and also is subsequently scattered as dust.

"Examination of the air for bacteria capable of development has shown that they are not suspended separately in the air, but that they dry on the surface of objects and do not enter the air until the dried mass breaks up, or unless the object on which the dried fluid rests is itself so light as to be carried away by the slightest breath of air. Such readily distributed carriers are particles of dust, consisting of bits of vegetable fiber, animal hair, epidermis scales, and such like. Hence we have to fear chiefly the soiling with phthisical sputum of materials consisting of vegetable products or animal hair, such as bed-linen, coverlets, clothes, handkerchiefs. Sputum that has dried in spittoons or on the floor gets detached only in larger pieces which do not readily float in the air. On the other hand, one can hardly imagine a more favorable contrivance for the distribution of the sputum as dust than that of allowing it to dry rapidly on stuff-garments, from which at each movement fibers fly off and carry the infective material into the air, where they remain suspended for some time, and when at last they fall to the ground the particles are easily caught up again by the slightest breath of air. The examinations of air undertaken by Hesse are very instructive on this point, and confirm fully what I have just stated."

"The following case is of great importance in connection with the etiology of local tuberculosis. It is narrated by Dr. E. A. Tscherning, in the 'Fortschritte des Medicin,' vol. iii, No. 3, 1885:*

"Maria P., aged twenty-four years, cook in the house of the late Prof. H., is of a completely healthy and strong constitution. She

* "Recent Essays by Various Authors on Bacteria in Relation to Disease." Selected and edited by W. Watson Cheyne, M. B., F. R. C. S. New Sydenham Society, London, 1886, *loc. cit.*

has never been affected with any scrofulous or tubercular disease. There is not the slightest trace of hereditary predisposition to tuberculosis in her family.

"Prof. H. died at the end of July (1884), from acute phthisis, which had lasted five or six months. His sputum toward the end of his life was almost a pure cultivation of tubercle-bacilli in pus. A few days before the professor's death, the patient wounded the palmar side of the first phalanx of her middle finger with a fragment of a broken vessel containing sputum. I first saw her fourteen days after the accident, when there were signs of commencing whitlow. For eight days carbolic fomentations were employed, with subsidence of the symptoms without suppuration; but a small nodule about the size of half a pea could be felt in the subcutaneous tissue. During the next few weeks the nodule remained somewhat painful and the tissue around was œdematous. At the end of August I made an incision and removed by means of a sharp spoon a small granulation-tumor scarcely as large as a pea, which lay between the skin and the sheath of the tendon. Healing took place in eight days by first intention under a dressing of iodoform and corrosive sublimate. For the time, improvement occurred; but when I saw the patient, at the beginning of October, she complained of pain on bending the finger. The skin and subcutaneous tissue were slightly swollen, and also the palm of the hand close to the phalanx. I could not make out any circumscribed swelling of the sheath of the tendon. By the advice of Prof. Studsgaard, she used local vapor-baths for some weeks, and presented herself again in the middle of November. We could then feel, through the somewhat œdematous skin, a very distinct swelling of the sheath of the flexor tendon. The movements of the finger were interfered with and there was considerable pain and tenderness. There were also two swollen glands above the elbow and two in the axilla. In other respects the patient was well. There was no trace of lung-disease.

"On November 21st Prof. Studsgaard removed the swollen axillary and cubital glands and amputated the middle finger at the metacarpo-phalangeal articulation, splitting up the palm and removing the tendon with its sheath as far as the middle of the palm. The subcutaneous granulation tissue was widely removed and scraped out. The operation was performed with the usual antiseptic precautions (corrosive sublimate, 1 per 1,000), and dressed with a sublimate wool and gauze dressing. The wounds healed by first intention in eleven days. The patient was well when discharged.

"The following pathological changes were found: The sheath of the tendon was filled with granulation tissue, its wall thickened. In the serous covering of the tendon were petechial hæmorrhages; but there was no pus nor caseous masses. There was no affection of the joints or bones. The granulation tissue showed under the microscope (after hardening in alcohol and staining with picro-carmine) very numerous young tubercles, with in many cases central caseous degeneration, frequently large cells and very beautiful giant cells often in the center of the nodules.

"The extirpated glands presented to the naked eye the appearance of simple hyperplasia without pus or caseous deposits. Under the microscope I found large numbers of large cells, with here and there small tubercles. Both in the granulation tissue and in the lymphatic glands I found in all the sections, stained by Ehrlich's method, well-marked tubercle-bacilli partly in the large cells, more especially in the giant cells, and partly at the margin of the necrotic places. The bacilli, as a rule, lay singly; but here and there there were two or three together, especially in V-shaped groups. Many of them contained the so-called spores.

"I saw the patient again at the end of January, 1885. She was quite well, and without chest-symptoms. No fresh lymphatic enlargements; linear scars with little tenderness; no sign of spread of the disease, either locally or generally.

"The microscopic appearances described here correspond entirely to what I have formerly found in about thirty cases of surgical tuberculosis (joints, sheaths of tendons, spondylitis, pyogenic membranes, lymphatic glands, testicles, tongue, and tuberculous kidneys), which I investigated from this point of view, formerly as prosector, and lately as assistant physician."

"In a communication made to the Académie des Sciences, by MM. Spillman and Haushalter,* and recorded in 'La Semaine Médicale,' the question of the spread of the tubercle-bacillus by means of the common house-fly is considered. The authors state that they have seen flies enter the spittoons containing the sputum of phthisical patients; they were then caught and placed

* London "Lancet," September 10, 1887.

in a bell-jar. On the following day several of these were dead. Examination of the abdominal contents and the excrement of these flies on the inside of the jar showed the presence of many tubercle-bacilli. The authors point out the wide dissemination of the disease which may take place in this way, and recommend as a preventive the employment of covers with a small opening."

"In an article in 'La Clinique,' on the 'Contagiosity of Tuberculosis,' by MM. Destrée and Slosse,* it is stated that from inquiries and observations made in Dr. Desmeth's wards during the present year it was found that, of fifty patients suffering from tuberculosis, contagion could be regarded as an undoubted etiological factor in twelve, heredity in thirty, while no cause could be traced in the remaining eight. The two factors, the authors say, are frequently present, and it is difficult to determine to which of them the preponderating influence is to be ascribed."

"By a number of carefully conducted experiments,† M. Cadéac and M. Mullet have undertaken to determine if the air expired by patients suffering with pulmonary tuberculosis can produce the disease by inhalation or by inoculation. They publish an interesting account of their experiments in a recent number of the 'Revue de Médecine.' The method which they adopted in the first series of experiments was as follows: A caoutchouc bag,

---

* London "Lancet," November 5, 1887.

† "New York Medical Journal," editorial, November 5, 1887.

having a capacity of from forty-five to fifty quarts, provided with a stop-cock, was partially filled by being breathed into by a patient in an advanced stage of phthisis. It was then filled to its utmost capacity with pure air, and in that way a vitiated atmosphere was created like that which is usually to be found in a phthisical patient's room. Rabbits, the susceptibility of which to tuberculosis is well known, were made to breathe in the air contained in the bag for an hour or two hours every day, by means of a certain muzzle, fashioned after the mouth-piece of Paul Bert's anæsthesia-inhaler. This was repeated for several days, and the three rabbits upon which the experiments had been performed were killed after the lapse of from twenty to forty days, and their organs examined. The results were entirely negative; all the viscera were found perfectly healthy. In another series of experiments, rabbits affected with catarrhal bronchitis were treated in the same way, but in them also the results were negative.

"The objection might be raised that in these experiments the degree of infection of the mixed air was variable, and the exposure too limited in duration. To exclude this source of fallacy, a third series of experiments was undertaken. A small box was divided into two compartments, in such a way that animals put into the two compartments had to breathe the same air, but could not come into bodily contact. Rabbits in which artificial tuberculosis had been produced were put into one of the compartments, and Guinea-pigs affected with induced bronchitis were placed in the other. Thus, for periods of two or three months, the non-tuberculous Guinea-pigs had to inspire the emanations from nineteen tuberculous rabbits; yet not one of the Guinea-pigs showed any signs of tuberculosis, either during its life or after being killed. In a fourth series of experiments, the vapor contained in the expired air of phthisical patients was condensed, and the resulting liquid was injected under the skin in rabbits, with antiseptic precautions. Twelve rabbits were thus treated, and the results were negative in all, with a single exception. In the rabbit that proved the exception two gray granulations, of the size of a pin's head, were found in the left lung. Thirteen Guinea-pigs were treated in the same way, and in none of them did infection take place. The authors pushed their experiments still further. There were two wards in the hospital, one of which was 120 feet long, 24 feet wide, and 21 feet high, containing about forty consumptive patients, and the other was 105 feet long, 18 feet wide, and of the same height as the first, having about thirty patients. The majority of the patients expectorated freely, and no precautions were taken with regard to the sputa. Several of the patients were in such an advanced stage of phthisis that they had to keep their beds constantly. Air collected from these wards, quite close to the beds occupied by patients, was subjected to condensation of its aqueous vapor, which was used in the same way as in one of the other series of experiments. Twelve Guinea-pigs were employed in these experiments, and the results were negative in all but two of them. In one of these two, a few small, gray granulations, containing tubercle-bacilli, were found in one of the lungs; in the other there were numerous foci of tubercular granulations in both lungs, and in some of the glands of the body.

"Although these experiments do not entirely disprove the infectiousness of tuberculosis, it must be said that they afford strong presumptive evidence of the innocuousness of the air expired by consumptive patients. The two instances of positive results obtained in the last series of experiments emphasize the necessity of disinfecting the sputa of phthisical patients, whether in a large hospital ward or in small apartments occupied by the sick and the healthy."

Dr. Cornet* publishes in the "Internationale klinische Rundschau" an account of a series of experimental investigations on tuberculosis which he has been conducting for the last two years in the Berlin Hygienic Institute. The experiments were divided into three groups: the first of these dealing with the air and dust in dwelling-houses, hospitals, etc.; the second group comprising observations directed to the parts of the body affected by tubercle artificially introduced in different situations; and the third group consisting of attempts to solve the problem of the possibility of rendering the tissues unsuitable as a cultivating medium for the tubercle-bacilli. In order to examine the walls and floors of rooms, the surfaces were washed over with sterilized sponges, which were then used to inoculate broth, the resulting culture

* London "Lancet," May 19, 1888, *loc. cit.*

being injected into the abdominal walls of three Guinea-pigs. The animals (if they did not die of some intercurrent affection) were killed forty days later and a careful necropsy made. Twenty-one hospital wards, in which most of the patients were phthisical, were examined in this way, the result being that from the dust of fifteen of them tuberculosis was set up. Similar observations made in lunatic asylums showed that the walls of these establishments are very frequently infected with tubercle. Private houses where persons affected with phthisis had lived gave likewise very distinct positive results; out-patient departments and surgical wards appeared, on the other hand, to harbor no tubercle. One important observation made was that, where phthisical people had been in the habit of expectorating on the floor, this was certain to yield infectious cultures; whereas, in cases where handkerchiefs or spittoons had always been used, the liability of the dust to prove infectious was very greatly diminished. Regarding the organs affected, Dr. Cornet fully confirms Koch's observation that, except the actual point of introduction, the organs most affected are the nearest lymphatic glands. Thus, when inhalation is the mode of infection adopted, the bronchial glands are the organs most affected; when injections are made into the abdominal walls, the inguinal glands of the side selected show the greatest degree of tuberculous infection; and, when the virus is introduced directly into the abdominal cavity, the omentum is the part most affected. The therapeutical observations were made with tannin, "pinguin," sulphureted-hydrogen water, menthol, corrosive sublimate, creolin, and creosote, all of which were given in much larger doses in relation to the body weight than any one would think of prescribing for human beings. In the case of corrosive sublimate, toxic symptoms were induced before the tubercle was injected, and care was taken with all the other remedies that the system was well saturated with them. Notwithstanding all this, however, every animal died, not the slightest hindrance being apparently caused to the development of the tubercle-bacilli by any of the remedies.

Dr. Cornet remarks that of course the results of experiments on Guinea-pigs must not be taken as necessarily holding good for human subjects, as he has himself repeatedly proved the great value of creosote in the treatment of phthisis. Again, some infected Guinea-pigs were sent to Davos, others being kept in Berlin, the conditions of life of the two sets of animals being rendered as similar as possible. All of them died in about the same time, and no perceptible difference was found in the degree of tuberculous infection of the tissues in the two classes of cases.

Brown-Séquard, Stokes, and others, claim that certain of the lower animals do not become tuberculous after inoculation with phthisical sputum if they are permitted to enjoy out-door life in pure air and are supplied with abundant nutriment, and that they rapidly become tuberculous after inoculation if these hygienic influences are denied them. Trudow's experiments, in which pure cultures of the tubercle bacillus were used for inoculation, corroborate these statements.

Councilman * states: "That tuberculosis is an infectious disease, and that the tubercle-bacilli are the infecting agents, are facts too well established to need further discussion. Probably the greatest advances in our knowledge of the disease that have been made since Koch's discovery of the bacillus have been as to the modes of infection; and this has been, not so much the question of the infection of the individual, but how further infection of the different tissues and organs takes place after the disease has established itself. It is especially to the work of Weigert that the most of our knowledge on this subject is due. We know now that the extent and character of the lesions produced are for the most part due to the number of bacilli which enter a part and to the manner in which they enter. We have the acute miliary tuberculosis, presenting all the clinical features of an acute infectious disease, which is due to general blood-infection by large numbers of bacilli entering the blood-current; then the general disseminated tuberculosis, in which there is also a general infection of the blood (but the bacilli which enter the blood are fewer, or they enter it at intervals); then the tuberculous pneumonia, affecting larger or smaller portions of the lung parenchyma, and due to the aspiration of large numbers of bacilli furnished by a focus of the disease, usually a cavity, in some other part of the lung. The miliary tubercles are most

* An address on "Predisposition in Tuberculosis," delivered before the Society of the Alumni of Bellevue Hospital, New York, April 4, 1888, by William T. Councilman, M. D. Published in the "New York Medical Journal" for April 21, 1888.

probably due to a comparatively small number of bacilli carried to a part by means of the blood-vessels or lymph-vessels.

"These are only some of the various lesions which may depend on the manner in which the bacilli enter a part and their number. The ways along which the bacilli can enter a part can be only three: they can enter it only by means of canals or ducts which penetrate it, or by the blood or lymphatic vessels.

"Infection of the individual can take place by means of the respiratory system, the alimentary canal, the genito-urinary tract, and the outer surface. Infection by the respiratory tract is the most common, and, next to this, infection by the alimentary canal, this being more common in children than in adults. The genito-urinary tract furnishes the route of infection in a much smaller number of cases, and a few instances have been known of infection from the outer surface of the body. It is not always easy to ascertain with certainty the primary lesion of the disease. This is often of so little moment anatomically, in comparison with the secondary lesions, that it may readily be overlooked, or placed in the same category with these.

"The epithelial covering of the surface of the body forms so perfect a protecting mantle that it is extremely improbable that infection of this ever takes place without a lesion. The bacilli may, however, enter the body by means of the alimentary canal without causing any lesion at the place of entry. This mode of infection is seen in the mesenteric phthisis of children, in which the mesenteric glands are the seat of an extensive tuberculous inflammation, often with no lesion of the intestine, or the bacilli may pass through the epithelium of the mouth or pharynx and the first seat of the disease be in the cervical lymph-glands. We know that the bacilli do pass through the epithelial covering of the small intestine, and in cases of intestinal tuberculosis the first formation of tubercles is in the lymphatic tissue beneath this.

"They may pass into the lymphatics without producing any lesions at the place of entry, and first make their presence known by a tuberculous inflammation of the glands. The anatomical structure of an organ must also influence the character and the extent of the tuberculous lesions produced in it. Thus the lungs, by means of the branching system of the bronchi into which the bacilli soon enter, favor the conveyance of the virus from one point to another—more than the liver, for example.

"Not only do the lungs favor a primary infection, but, when once the disease is established, further infection of other parts easily takes place."

Predisposition to Phthisis.—Hirsch, in considering congenital or acquired predisposition to phthisis, says that "the disposition must be assigned exclusively to abnormal states of the respiratory organs themselves, which had either been congenital or been called forth by external influences acting upon the lungs directly or indirectly. In the same class of directly noxious things predisposing to an attack of phthisis, we have to reckon passing the time in crowded and ill-ventilated places, the air of which is laden with organic decomposition-products or minute particles of mechanical or chemical irritants; also, chronic bronchial catarrh (depending sometimes upon the causes just mentioned, and sometimes upon the weather) and chronic pneumonias (or pneumonias that had not cleared up), particularly broncho-pneumonias.

"Among the influences operating indirectly we have to include all those things that are detrimental to the nutrition of the organism in general, making it vulnerable through defective repair of the tissues; and that vulnerability, so far as concerns the lungs, is the cause of their predisposition to morbid processes in general and to consumption in particular. This is the explanation of the disease being so notoriously common among persons living in miserable circumstances and reduced to struggle with want and cares, also of its development in those who have been entirely worn out and reduced by severe sickness.

"On the other side, we may thus explain the exemption from phthisis of many parts of the world by reason of their favorable weather-conditions and the consequent rarity of all pulmonary affections therein. The immunity from consumption enjoyed by the natives of elevated regions seems to me to be referable to a peculiarly strong development of their breathing-organs and a corresponding power of resistance in them to noxious influences from without. It is proved that this is not at all an affair of 'purity of the atmosphere,' as some have supposed, by the fact that the state of hygiene in the towns of Ecuador, Bolivia, and Peru, situated at great elevations, is by no means distinguished for its excellence, for cleanliness in the houses and streets, adequate ventilation of rooms, and the like."

Koch states that "although a good many of the phenomena classed together under the head of 'predisposition' may be referred to simple and easily explained conditions, some facts remain difficult or impossible to interpret, compelling us for the present to accept the view of a varying liability. Most important of all is the striking difference between the course of tuberculosis in children and in adults, and again the undeniable predisposition to tuberculosis that exists in some families.

"In the latter instance, many of the cases of illness where predisposition is supposed to be an important factor might rather be ascribed to increased opportunities of infection, and there are also peculiar predisposing influences connected with the family constitution to be thought of, such as a tendency to catarrhal affections of the respiratory organs and imperfect development of the thorax. Still, many carefully observed cases remain which can not be explained in this manner. Further, individual cases of the disease have often shown that a person is not at all times an equally favorable subject for the development of the parasites, for it not infrequently happens that tubercular foci which had reached a fair size contract, cicatrize, and heal up. That means, however, that the same body which afforded a suitable soil for the tubercle-bacilli on their first invasion, so that they were able to multiply and spread, has by degrees lost the qualities favorable to them and has changed into a bad soil, thus preventing further growth of the bacilli; so that in the same person there was at one time a liability to tuberculosis and again at another time none. Further investigation is required to show what occasions this difference, whether it is due to a change in the chemical composition of the juices of the tissues or to physical conditions. So much is certain, that these differences exist; and there is nothing against the view that conditions similarly favorable or unfavorable to the tubercle-bacilli are present in some human beings not only temporarily, but for the whole lifetime."

Leichtenstern, speaking of Koch's brilliant discovery, says, "there are still many lacunæ and open questions facing us in the etiology of consumption, and these offer to the practitioner, to the statistician, to the pathologist who studies the history and geography, to the pathologist who experiments, and to the bacteriologist, a wide field wherein to co-operate. It is not by the power of any sudden enthusiasm, treating the infective nature of phthisis as if it were already made perfectly clear, that the new doctrine will be made secure of its position, but by earnest work and prolonged study."

VII.

CONCLUSIONS.

RINDFLEISCH says: "Any one who wishes to give a special account of tuberculosis of the lungs must necessarily commence by stating his views on the subject in general; for so much has been written, and so many different opinions prevail concerning tubercle, that, unless an author first states to what condition he applies this word, he can not be understood."

In the first chapter of this work I have attempted to record the opinions of prominent writers on pulmonary consumption, from the time of Hippocrates up to the present day, believing that a knowledge of their views, and of the points upon which they have failed to agree, will give a better understanding of the subject than may be otherwise obtained.

The statistical data given in the foregoing pages of this work are, as I have previously observed, the most accurate attainable at the present time.

The climatological study of the future must be based, as Gihon* has stated, "upon rational methods of investigation. The mere recording of meteorological fac-

---

* The opening address of the president, Albert L. Gihon, M. D., U. S. N., of the "Section in Medical Climatology and Demography." "International Medical Congress," Washington, D. C., September 5, 1887.

tors is not sufficient. Determinate climatic characters are not easy to formulate. . . . There are few specific climatic diseases. Local conditions of insanitation are more responsible for the production of diseases than the general influences of climate. By appropriate regulations of habits, clothing, and diet, the morbific effects of climate may be modified or averted, or its sanitary or therapeutic influence heightened.

"The data for future generalizations must be furnished by accurate and laborious collective investigation. Vital statistics must in future be something more than mere records of so many deaths, births, or marriages. Morbility records must form the principal data for the vital statistics of the future. To have these records accurate, voluntary effort can not be depended upon; they must be made under governmental authority, if our vital statistics shall serve as the basis of trustworthy generalizations."

In the tenth "United States Census Mortality and Vital Statistics Reports," concerning which Pepper* says: "Nor can I neglect this opportunity of referring to the great practical value of this colossal work. Despite the serious defects of the statistics resulting from the absence of any national system of registration of vital statistics, *such as is relied upon by all other civilized nations*† for the purpose of ascertaining the actual

---

* "A Contribution to the Climatological Study of Consumption in Pennsylvania." By William Pepper, M. D., LL. D., Philadelphia. Published in the "New York Medical Journal" for December 4, 11, and 18, 1886.

† The italics are mine. G. A. E.

movement of population, the improved method employed in this tenth census, and the ability shown by Dr. Billings in the arrangement and analysis of the results, render the two volumes which have just appeared highly valuable to the profession, and highly creditable to the genius and energy of their distinguished author." Billings states: "While the original schedules of deaths contain data from which it would be possible to make, in part at least, the necessary deductions to express the true tendency to this disease (consumption) in these (certain) localities, such calculations have been made impossible from the *want of clerical force*." *

As Flick truly observes: "Medical science has grown beyond the mere art of prescribing remedies; it has become a science of protecting man against disease, and enabling him to attain his threescore and ten. As government exists for the good of society, it ought to avail itself more extensively of so powerful a means to its end."

Concerning the conclusions which may be deduced from the evidence which has been submitted in regard to the geographical distribution of phthisis, I can not do better than give the following brief summary of Hirsch's conclusions: †

"Phthisis is everywhere prevalent, but it is rare in polar regions, and rarer still at great altitudes. The main factor in its production is over-crowding and bad hygiene. Heat and cold, *per se*, have no influence.

* The italics are mine. G. A. E.

† "British Medical Journal."

Damp, when conjoined with frequent oscillations of temperature, predisposes to the disease; but humidity of the air is less important than dampness of soil. Occupation is extremely important, but mainly indirectly, as tending to good or bad hygienic conditions."

With reference to the part played by the tubercle bacillus, it is reasonable to believe that it holds the same etiological relation to pulmonary phthisis that certain other micro-organisms hold to external surgical affections, to septic diseases of the (post-partum) uterus or its contiguous tissues, etc.

That pulmonary phthisis occasionally terminates in recovery there can be no doubt. Cases are frequently reported by competent observers in which recoveries have taken place. The following recent report of a case of spontaneous recovery from pulmonary consumption is of so much interest in this connection that I give it unabridged:*

"The subject of this communication, a young woman aged twenty-three, first came under my observation in the latter part of 1884. Her family history was bad; her father and a brother had succumbed to lung-disease; a sister was lying ill of phthisis, of which she shortly died; and she herself was weak, anæmic, and very dyspeptic.

"Notwithstanding these hindrances, she managed, with the aid of arsenic and iron, to fulfill her arduous duties as a teacher, with but slight intermissions, until August, 1886, when increasing weakness, anorexia, and scrofulous inflammation of the right cervical glands laid her aside. An examination of the lungs at this time revealed nothing very definite. In due time a large quantity of

---

* "Recovery from Subacute Phthisis." By A. G. Auld, M. D., London "Lancet," February 11, 1888.

characteristic pus was twice abstracted from the neck. Shortly afterward, about Christmas, an intense inflammation attacked the great joint of the thumb of the right hand, leading to sinuses, discharging evidently tubercular pus. After two months this discharge, with its attendant phenomena, began to disappear, while simultaneously were developed the signs and symptoms of phthisis pulmonalis. The disease at first threatened a somewhat severe course. The left lung became extensively involved; temperature ranged from 102° to 104°; cough and expectoration considerable; night-sweats profuse; circulation very weak; appetite gone. 'Those are the gloomiest cases of phthisis,' says Dr. Sutton, 'where there are anæmia and weakness of the pulse.'

"In the middle of May, Bergeon's treatment, then attracting attention, was contemplated, but the patient was considered too weak, and the idea was abandoned. About this time, however, the stomach, which hitherto had resisted all treatment, began to show signs of improvement, and as much suitable food as could be borne was administered. Occasional attacks of sickness were best relieved by a few drops of solution of cocaine, with ice, and the bowels were kept well open. No antiseptics were employed, as I have been invariably disappointed in their use. This improvement steadily increased, accompanied by a very pronounced amelioration of the general symptoms and a very remarkable gain of flesh, till by the middle of July the pyrexia and night-sweats had almost entirely ceased. The expectoration nevertheless continued, and there were the physical signs of a vomica in the left apex, while in the right moist râles were audible. Considering the case unusual, I communicated with Prof. Hamilton, of Aberdeen, in the month of August, who kindly examined the sputum, and reported that, after a very careful examination, he found that 'it contained the tubercle bacillus in considerable abundance.' From this period onward the patient was rapidly recovering, and by the end of September the cough and expectoration had nearly ceased, the moist sounds had almost vanished, and vesicular breathing was partially restored over the damaged areas. Early in October menstruation took place for the first time for several years, the discharge being normal as to duration and amount. By the beginning of November the cough had entirely disappeared, no moist râles were audible, menstruation had reappeared, and hitherto this favorable condition is maintained.

"REMARKS.—I think it may be fairly asserted that this is a very striking case of cured phthisis, considering that the patient was housed on a damp soil, in comparatively unfavorable surroundings, and recovered without the use of any of the 'special' means of cure."

The generally recognized fact that pulmonary phthisis depends upon impure air more than upon any other etiological factor for its origin, has led to the aseptic (climatic) air treatment for its relief. That the best results have been secured at great altitudes admits of no doubt. The influences supposed to be productive of the beneficial results reported are, aseptic air, attenuated air, dry air, cold (tonic) air, and ozone.

It has been shown, in a previous chapter of this work, that respiration in attenuated air demands greater respiratory energy than it does in an atmosphere of sea-level tension in order to get the same (a sufficient) quantity of air into the lungs. Respiration of attenuated air produces positive mechanical effects of undoubted benefit, not only upon the pulmonary organs *per se*, but also upon the circulation, by diminishing the atmospheric resistance to the passage of blood through the lungs and tissues generally, or, in other words, by lowering the arterial pressure.

The following reports and discussions on the "Treatment of Consumption by Residence at Great Altitudes" are of much interest in this connection:*

"Dr. C. Theodore Williams read a paper on the 'Results of the Treatment of Pulmonary Consumption by Residence at High Altitudes,' as exemplified by an analysis of 141 cases, of which the fol-

* "Royal Medical and Chirurgical Society," May 8, 1888. Published in the London "Lancet" for May 12, 1888.

lowing is an abstract: The author offers a contribution from his own practice of 141 cases of phthisis treated in sanitaria varying in altitude from 5,000 to 9,000 feet, in the Alps, the Rocky Mountains, and the South African Highlands, during the last nine years, in order to deduce certain practical rules therefrom. The 141 cases have been tabulated for statistical purposes under the following headings: Sex, age, length of illness before the commencement of mountain residence, hæmoptysis, history and nature of cases, state of the lungs, medicine and diet, length of residence at high altitudes. The Alpine climate is then compared with that of Colorado and the South African Highlands. The results of this treatment have been tabulated under the heads of *general*, referring to the general health, vigor, and weight, and *local*, including the conclusions arrived at from the examination of the lungs. The general results are divided into (1) cured, 41·13 per cent, where the restoration to health was complete; (2) greatly improved, 29·78 per cent; (3) improved, 11·34 per cent; (4) deteriorated, 17·02 per cent; thus giving a total of 82·25 per cent improved and 11·34 per cent deteriorated, including 13·47 per cent of deaths. The local results of the 141 cases yield improvement greater or less in 74·82 per cent (including arrest in nearly 44 per cent), deterioration in 21½ per cent, and a stationary condition in 3·59 per cent. Among the first-stage cases there was improvement in 91 per cent, and arrest of disease in 63 per cent, with deterioration in nearly 7½ per cent. Cases of unilateral first stage give 92 per cent improved and 70½ per cent of arrests, and cases of bilateral affection yield 87·09 per cent of improved and 48·38 per cent of arrests. In the second- and third-stage cases there was improvement to a greater or less extent in 46 per cent, arrest in 10 per cent, and deterioration in 46 per cent. Single-cavity cases gave better results than cavity cases with the opposite lung involved, and left-lung cavities showed a less tendency to change, either for better or worse, than the right-lung ones. The following conclusions are arrived at: 1. That prolonged residence at high altitudes produces great improvement in the majority of consumptive patients and complete arrest of the disease in a considerable proportion, such arrest being in a more or less degree permanent. 2. That in order to secure these advantages patients must be free from pyrexia and all acute symptoms, and must possess sufficient lung-surface to adequately carry on the process of respiration in the rarefied atmosphere. 3. That the influence of the climate seems to promote a change in the lungs, either of a curative or destructive character, and to oppose quiescence. 4. That residence at high altitudes causes enlargement of the thorax, hypertrophy of the healthy lung-tissue, and the development of pulmonary emphysema around the tubercular lesions, and that this expansion of the chest is accompanied by diminution of the pulse and respiration rate. 5. That it is probable that the arrest of consumptive disease is partly owing to the pressure exercised on the tubercular masses by the increasing bulk of the surrounding tissue. 6. That the above local changes are accompanied by general improvement shown in the cessation of all symptoms, and the gain of weight, color, and of muscular, respiratory, and circulatory power. 7. That consumptives of both sexes benefit equally by mountain residence, but that the age of the patient exercises considerable influence on the result. 8. That the high-altitude treatment seems to be specially adapted in cases where heredity and family predisposition are present. 9. The climate is useful in cases of hæmorrhagic phthisis, and that hæmoptysis is of rare occurrence at the mountain stations. 10. That mountain climates are most effective in arresting phthisis where the disease is of recent date; but they are also beneficial in cases of longer standing. 11. That the special effects of high-altitude residence on the healthy and sick are common to all mountain-ranges of elevations of 5,000 feet and upward. 12. That to insure the full advantages of high-altitude residence, a period of at least six months is necessary in the majority of consumptives. In cases of long standing and extensive lesions, one or two years are often requisite to produce arrest of the disease. 13. That, in addition to the above examples, mountain climates are beneficial in (1) cases of imperfect thoracic and pulmonary development; (2) chronic pneumonia without bronchiectasis; (3) chronic pleurisy, where the lung does not expand after removal of the fluid; (4) spasmodic asthma, without much emphysema; and (5) anæmia. 14. That they are contra-indicated in the following conditions: (1) Phthisis with double cavities, with or without pyrexia; (2) cases of phthisis where the pulmonary area at low levels hardly suffices for respiratory purposes; (3) catarrhal phthisis; (4) erethitic phthisis, or phthisis where there is great irritability of the nervous system; (5) emphysema; (6) chronic bronchitis and bronchiectasis; (7) diseases of the heart and greater vessels; (8) affections of the brain and spinal cord, and conditions of hypersensibility of the nervous system; and (9) where the patients are of advanced age and where they are too feeble to take exercise.

"Dr. Bowles discussed the pathology of sunburning, and thought that the causes of it had some share in the improvement of phthisical patients. He argued that the reflected light from snow was a potent cause of sunburning, far more important than heat-rays or the atmosphere. The ultra-violet rays with most chemical action probably were chiefly causative of it. He found that painting the face with a brown pigment prevented sunburning.

"Dr. Hermann Weber gave a few of the results of his own statistics of 106 cases treated at high altitudes. Of these, 38 were cured, 42 improved, 16 remained stationary, and 10 deteriorated. Of 70 in the first stage, 36 were cured, 28 improved, 11 remained stationary, and 6 grew worse. Of 32 in the second stage, 2 were cured, 13 improved, 11 remained stationary, and 6 deteriorated. Of 4 in the third stage, 1 was improved, 1 remained stationary, and 2 deteriorated. While granting that high altitudes did thus effect great good, Dr. Weber pointed out that equally good results might be obtained at lower levels if systematic medical and dietetic treatment were carried out, as at Falkenstein, near Frankfort. From this place, which is only about 2,000 feet above the sea-level, Dettweiler's statistics (which are perfectly reliable) prove this contention. In Dr. Weber's cases the gain in weight was observed in 58 cases, weight remained stationary in 40, and 8 lost flesh. It was remarked, however, that patients could be well nourished before being sent to high altitudes, so that a gain in weight might thus be anticipated by previous treatment.

"Dr. De Haviland Hall asked for experience in cases of laryngeal phthisis treated at high altitudes. In his experience such cases did badly.

"Dr. Pollock criticised Dr. Williams's paper, and maintained that equally good results could be, and in fact were, obtained by medical and hygienic treatment in London. He passed each stage of phthisis in review, and pointed out that it was precisely those cases which were benefited at high altitudes that did well at home. Cases of congestion and fever were always the worst, and it was well known that high altitudes did not suit such.

"Dr. Tucker Wise maintained that high altitudes were most beneficial in phthisis, and that the improvement was far greater than could be obtained elsewhere. The mode of action of high altitudes was discussed and a high place given to the asepticity of the atmosphere.

"Dr. Ewart supported the conclusion of Dr. Williams's paper, and laid special stress on the immense benefit that accrued from the sudden change to a place of hopefulness, cheerfulness, and rest. He contrasted the cheerfulness and *bien-être* of the patients at high altitudes with those at lower levels.

"Dr. Huggard contended that high altitudes, on the whole, gave the most satisfactory results. He attributed the beneficial effects largely to the rarity of the atmosphere and its effects on the human organism.

"Paul Bert's experiments on the influence of low tension on the escape of gases were applied to illustrate the effects of the treatment at Davos and elsewhere.

"Arterial tension was lowered, and this might explain the freedom from hæmoptysis. When hæmoptysis did occur, it was owing generally to too sudden arrival at great heights. The hæmorrhage was then of the order known to balloonists.

"Dr. Quain believed most firmly that as much good could be obtained at home or at low levels. The constitutional state was the most important factor in bringing about recovery.

"Dr. Williams in reply argued that, though good results could be obtained at home or on sea-voyages or in warm places, high altitudes yielded far better results. This was shown in the actual arrest of the disease, which did not happen, or very rarely, when cases were treated elsewhere. By arrest he meant total disappearance of all physical signs of lung-mischief as well as restoration of the constitutional state. In his experience, laryngeal phthisis was also a contra indication for the high-altitude treatment."

Rhazes wrote, nearly one thousand years ago, that patients die from consumption because the lungs can not be treated like external parts.

Loomis,* speaking of the climatic treatment of consumption, has said: "The air must be pure, aseptic if you choose, which could be taken in the lungs on the same principle that antiseptics were used externally in the treatment of surgical affections. Cavities in the

* "New York Academy of Medicine," October 20, 1887. Published in the "New York Medical Journal" for October 29, 1887.

lungs could not be washed out with antiseptic solutions, and he doubted whether such solutions could be applied by inhalations, so as to destroy the cause of the morbid processes going on in the lungs; but, if the lungs could be bathed constantly with aseptic air, all was done that could be in the way of local treatment of pulmonary phthisis."

Tyndale* says: "It is my earnest desire to draw attention to what seems to me the future path to be pursued in therapeutics, namely, to endeavor to bring nutrition to the highest point attainable, and retain it there long enough to enable us to pursue a *systematic course of antiseptic treatment of the general condition, laboring under chronic septicæmia, as well as of the local lesion.*"

The writer believes that the *respiration of antiseptic air by phthisical subjects* will be found, in the future, to be as successful in the treatment of consumption as topical antiseptic influences have been in the treatment of external surgical affections. Although antiseptic air does not exist in nature, medical science is able, by means of appliances now at command, to produce it, by combining aseptic air with appropriate antiseptic agents, in sufficient quantities, at least, for respiration by phthisical subjects during their sleeping hours. It is the hope of the writer, and doubtless of many others, that some such method of treatment for consumption, as well as for

---

* "Treatment of Consumption." By J. Hilgard Tyndale, M. D., New York, 1882.
The italics are mine. G. A. E.

other pulmonary diseases, may soon develop from what at present may seem to be only a Utopian idea.

NOTE.—In the "New York Medical Journal" for March 6, 1886, the writer describes a new instrument for the local antiseptic treatment of pulmonary phthisis.

This apparatus has been improved since that time, so as to permit of the simultaneous operation of compressed and artificially purified air, in conjunction with the topical application of stimulating and antiseptic medicament.

The results which have been obtained up to the present time by the writer, by means of this apparatus, are given in the following table:

STAGES.	Cases.	Improved.	Not improved.	Recovered.	Died.
Consolidation......	57	9	3	44	1
Softening.........	33	20	4	8	1
Excavation........	64	23	6	3	32
Total.........	154	52	13	55	34

It will be observed that these results are not so good as those reported in the article referred to. This is due in most part to the fact that a larger percentage of advanced cases have been treated since the first report was made, as a comparison of the tables will show.

The writer expects to make use of this instrument, in the near future, for the production of antiseptic atmospheres for continuous respiration by consumptives during their sleeping hours.

THE END.

A DICTIONARY OF MEDICINE, including General Pathology, General Therapeutics, Hygiene, and the Diseases peculiar to Women and Children. By Various Writers.

Edited by RICHARD QUAIN, M. D., F. R. S.,

Fellow of the Royal College of Physicians; Member of the Senate of the University of London; Member of the General Council of Medical Education and Registration; Consulting Physician to the Hospital for Consumption and Diseases of the Chest at Brompton, etc.

In one large 8vo volume of 1,834 pages, and 138 Illustrations. Half morocco, $8.00. Sold only by subscription.

This work is primarily a Dictionary of Medicine, in which the several diseases are fully discussed in alphabetical order. The description of each includes an account of its etiology and anatomical characters; its symptoms, course, duration, and termination; its diagnosis, prognosis, and, lastly, its treatment. General Pathology comprehends articles on the origin, characters, and nature of disease.

General Therapeutics includes articles on the several classes of remedies, their modes of action, and on the methods of their use. The articles devoted to the subject of Hygiene treat of the causes and prevention of disease, of the agencies and laws affecting public health, of the means of preserving the health of the individual, of the construction and management of hospitals, and of the nursing of the sick.

Lastly, the diseases peculiar to women and children are discussed under their respective headings, both in aggregate and in detail.

Among the leading contributors, whose names at once strike the reader as affording a guarantee of the value of their contributions are the following:

ALLBUTT, T. CLIFFORD, M. A., M. D.
BARNES, ROBERT, M. D.
BASTIAN, H. CHARLTON, M. A., M. D.
BINZ, CARL, M. D.
BRISTOWE, J. SYER, M. D.
BROWN-SÉQUARD, C. E., M. D., LL.D.
BRUNTON, T. LAUDER, M. D., D. Sc.
FAYRER, Sir JOSEPH, K. C. S. I., M. D., LL. D.
FOX, TILBURY, M. D.
GALTON, Captain DOUGLAS, R. E. (retired).
GOWERS, W. R., M. D.
GREENFIELD, W. S., M. D.
JENNER, Sir WILLIAM, Bart., K. C. B., M. D.
LEGG, J. WICKHAM, M. D.
NIGHTINGALE, FLORENCE.
PAGET, Sir JAMES, Bart.
PARKES, EDMUND A., M. D.
PAVY, F. W., M. D.
PLAYFAIR, W. S., M. D.
SIMON, JOHN, C. B., D. C. L.
THOMPSON, Sir HENRY.
WATERS, A. T. H., M. D.
WELLS, T. SPENCER.

"Not only is the work a Dictionary of Medicine in its fullest sense; but it is so encyclopedic in its scope that it may be considered a condensed review of the entire field of practical medicine. Each subject is marked up to date and contains in a nutshell the accumulated experience of the leading medical men of the day. As a volume for ready reference and careful study, it will be found of immense value to the general practitioner and student."—*Medical Record.*

New York: D. APPLETON & CO., 1, 3, & 5 Bond Street.

www.ingramcontent.com/pod-product-compliance
Lightning Source LLC
Chambersburg PA
CBHW021957220426
43663CB00007B/851
9783337187453